Composite Interfaces in Mechanical Design

Online at: https://doi.org/10.1088/978-0-7503-5688-6

Composite Interfaces in Mechanical Design

Parvez Alam
School of Engineering, Institute for Materials and Processes,
The University of Edinburgh, Edinburgh, United Kingdom

IOP Publishing, Bristol, UK

ISBN 978-0-7503-5688-6 (ebook)
ISBN 978-0-7503-5686-2 (print)
ISBN 978-0-7503-5689-3 (myPrint)
ISBN 978-0-7503-5687-9 (mobi)

DOI 10.1088/978-0-7503-5688-6

Version: 20241001

IOP ebooks

British Library Cataloguing-in-Publication Data: A catalogue record for this book is available from the British Library.

Published by IOP Publishing, wholly owned by The Institute of Physics, London

IOP Publishing, No.2 The Distillery, Glassfields, Avon Street, Bristol, BS2 0GR, UK

US Office: IOP Publishing, Inc., 190 North Independence Mall West, Suite 601, Philadelphia, PA 19106, USA

Mimi

Meem

Mia

Mymensingh

Mimsuli

Contents

Preface

My previous book, 'Composites Engineering: An A–Z Guide', also published by the Institute of Physics, was aimed at *arming inductees* in the world of composites with a reference text that they could use to essentially *learn the language* of composites engineers. My hope since then has been to create a series of books that would branch out further than the first in order to develop deeper insights into more specific topics. I have nevertheless decided to set some boundaries on the total length of each, to ensure that every book can feasibly be used by both educators and students as reading material in tertiary-level courses, whilst concurrently serving as a useful reference aid for researchers. This second book is thence the first of a series of natural follow-on books from my earlier book, and will delve more deeply into the subject area of 'Composite Interfaces in Mechanical Design'. It provides deeper insights into, and expands beyond, some of the interface-relevant A–Z topics of my first book on composites, whilst (hopefully) avoiding verbose explanations, or excessive levels of detail. It should be noted that, unlike the first book, the focus here is primarily on fibre-reinforced polymer (FRP) composites. This book is subdivided into nine chapters, each covering a different generic theme related to composites and their interfaces. While each chapter is to an extent self-contained, there are legitimate inter-chapter overlaps, and where these exist, I have indicated within the text where aligned or complimentary material can be found in another chapter or section.

Chapter 1 of this book is introductory, discussing initially the fibre–matrix interface and connecting the effects and behaviours of interfacial shear with context to the stress–strain profile of a unidirectional FRP composite in tension. The chapter then continues with a section on composite delamination, the effects of z-thickness effects, and on how joining can instigate delamination failure. The effects of geometry in delamination are also covered, as are delaminations owing to laminate connections and discontinuities. This is followed by an introductory section on the strain energy release rate for mode I and II fractures. These are important considerations when composites fail at their interfaces, and as such, these are discussed from a design perspective, taking into consideration inherent relationships between the two fracture modes. The chapter concludes by providing examples of interfacial failure in engineered composite structures, using the specific examples of helicopter rotor blades and rockets constructed using composite materials. A close-to-home survey case of interfacial failure is conducted on the remains of Darwin III, a composite rocket built by Endeavour, The University of Edinburgh's rocketry society, which failed on impact in Portugal in 2022.

Chapter 2 of this book considers the chemical characterisation of fibre interfaces by Fourier transform infrared (FTIR) spectroscopy. The chapter begins by introducing the method and continues by providing information on typical bands and peaks observed in the following fibres: carbon fibres (including sized and unsized); glass fibres (including sized and unsized); m- and p-aramid fibres; and surface-treated and untreated aramid fibres. The chapter concludes with a section on natural fibres and includes the FTIR characterisation of cellulose polymorphs and the effects of chemical treatments such as alkalisation (mercerisation) and acetylation on spectral peaks and bands.

Chapter 3 introduces the four mechanisms of bonding in fibre-reinforced composites including mechanical interlocking of adhesion, electrostatic adhesion, chemical adhesion, and diffusion adhesion including entanglements. The chapter also explains in detail the effects roughness has on adhesion and wetting, providing models for concave and convex wetting, as well as more well-known models of roughness such as the Wenzel and Cassie–Baxter models alongside lesser-known derivative models.

Chapter 4 detracts slightly from interfaces to provide further detail on the formation of interphases as they are essentially borne from interfaces. The chapter begins by explaining the physical nature of interphases and their formation into continuous and discontinuous phases within composites and also in terms of the measured thicknesses of interphases as reported in the literature. The chapter then continues to detail the kinetics of interphase formation, which is followed by a section on the thermodynamics of interphase formation. The chapter concludes with a discussion on the effects interphases may have on established composite models such as the Rule of Mixtures, the Transverse Rule of Mixtures, and modulus and strength models for nanocomposites.

Chapter 5 considers the surface treatment of reinforcing fibres. Details are provided on common carbon and glass fibre-sizing chemistries, the identification of sizing on fibre surfaces using IR spectroscopic methods, and on the reported effects of sizing on mechanical performance, both at a fibre/matrix level (in terms of interfacial shear strength) and at a composite level (in terms of interlaminar shear strength). Natural fibres are also discussed at the end of this chapter, and although there is a plethora of chemical, biological, and physical treatments applicable to natural fibres (for use as reinforcing in composites), this section details four chemical processes (alkalisation, acetylation, silane coupling agents, and maleic acid coupling agents), as well as one physical process (plasma treatment).

Chapter 6 discusses interface morphology, specifically in the cases of glass and carbon fibre reinforcements. The majority of fibre morphology studies in the composites community are based on roughness measurements. As such, the chapter focusses on roughness and begins by discussing common types of roughness parameters. The chapter continues to ascertain the notable differences in roughness for both sized and unsized glass and carbon fibres and provides reasons for these differences. Roughness is also plotted against mechanical properties in this chapter, as it is an area that has not been dealt with in great detail in the literature despite there often being correlations between the two that cannot be ignored.

Chapter 7 provides illustrations, information, and standards for different macroscale and microscale mechanical testing methods relevant to interfaces. Macroscale test methods include lap shear tests; double cantilever beam tests (for mode I); tapered double cantilever beam tests (for mode I); wedge-peel tests and impact wedge-peel tests (for mode I); end- notch fracture tests (for mode II); end-loaded split tests (for mode II); four-point end-notched flexure tests (for mode II); edge ring crack torsion tests (for mode III); and mixed mode flexure/bending tests (for the mixed modes I and II). Fibre/particulate microscale test methods include single-fibre pullout tests, micro-bond tests, micro-indentation tests, Broutman tests, and fibre fragmentation tests.

Chapter 8 begins by elucidating the fundamental principles of dynamic mechanical analysis (DMA) techniques, as applicable to fibre-reinforced composites. The methods for interpreting and extracting information on the viscoelastic properties of composites using DMA curves are subsequently discussed. Following this, useful DMA models for composites engineers are provided, including the adhesion factor, the degree of entanglement, and the reinforcing effectiveness factor.

Chapter 9 is the final chapter of the book and discusses fracture and failure in greater detail than has been covered in other areas of the book. The chapter focusses on the effects of tensile loading, fibre orientation, and cross-laminating on composite properties and behaviour. Types of fibre failure in tension are considered at the micromechanics scale, including possible interfacial failures through tensile loading, and the compressive modes of failure in fibre-reinforced plastic composites. The chapter concludes with a section on interlaminar shear failure in composites, highlighting relevant standards and deliberating on common modes of failure as reported in the literature.

Author biography

Parvez Alam

Parvez Alam is currently a Reader in Mechanical Engineering at The University of Edinburgh, UK. He is a Fellow of the Institute of Materials, Minerals and Mining, a Fellow of the Institution of Mechanical Engineers, a Fellow of the Royal Society of Biology, a Chartered Mechanical Engineer, Chair of the Natural Materials Group: a Technical Community of the Institute of Materials, Minerals and Mining, and an Adjunct Professor in the Faculty of Biology at Universitas Gadja Mada, Indonesia. He was awarded the 2012 Per Brahe Prize for his research in biomimetics, and became a Technology Academy of Finland Laureate for his work on coral-inspired crystal engineering in 2014, an award conferred by the Swedish Academy of Engineering Sciences in Finland.

Parvez earned his Doctorate from the University of Bath in 2004 under the supervision of Dr Martin Ansell FIMMM, where his research focussed on mechanically characterising timber-reinforcing and repair systems using bonded-in fibre-reinforced plastics. He then moved to the Department of Chemical Engineering at Abo Akademi University, Finland, where he worked with Professor Martti Toivakka in the areas of paper science, mass flow, micromechanics, paper coatings, and multiphysics modelling. In 2013, he was awarded the title of *Docent* (Habilitation) from Abo Akademi University in Finland and built a research group in Composite Materials and Biostructures. This new trajectory led to a wide range of new and interesting discoveries in the area of comparative biomechanics, examples of which include the biomechanical and kinematic characterisation of the tree-climbing, water-hopping fish (*Periophthalmus variabilis*) with colleagues from Universitas Gadjah Mada in Indonesia; the discovery of a spider species (*Nephilengys cruentata*) in Kenya that spins the world's toughest recorded egg sac silk; the role of microstructure in guiding the passive decorating strategies of the crab *Tiarinia cornigera*; and work into the biomechanics of sperm whale bone architectures with colleagues at the University of Cape Town in South Africa.

After moving to Scotland, he held the post of Marie-Curie Very Experienced Researcher at The University of Edinburgh between 2016 and 2018. During this tenure, he worked with Professor Conchúr Ó Brádaigh FREng FRSE on the fatigue modelling of carbon fibre reinforced composites for use in tidal turbine blades, and in addition, took a short sabbatical to travel to the Kalahari Desert in Namibia to learn Ju'Hoansi, one of the world's oldest extant click-languages. While in Namibia, he also seized the opportunity to research the composite designs and manufacturing methods of Ju'Hoansi arrows, work that has since been publicly presented, and published in learned society magazines. After a second short sabbatical to gain a Professional Certificate in Innovation and Technology at MIT, he began his tenure as an academic at The University of Edinburgh, where he currently runs a group that conducts research on a broad range of interdisciplinary topics related to biomimetic design, mechanical metamaterials, comparative biomechanics, robotics, multi-body systems, and composites engineering. Some research highlights from his group at Edinburgh include the development of 3D projected 4-polytopes as a new class of mechanical metamaterial; the morpho-mechanical characterisation of varanid lizard claw-gripping efficiencies; the release of AInsectID—an open source artificially intelligent software for insect identification; and the design of a walking necro-robot beetle with the highest payload-ratio of any walking robot recorded to date. At Edinburgh, he also lectures courses on engineering design, having previously developed courses in composites engineering, computational modelling, and biomimetics while in the Nordic and South East Asian regions.

IOP Publishing

Composite Interfaces in Mechanical Design

Parvez Alam

Chapter 1

An overview of composite interfaces

1.1 Introduction

This first introductory chapter is purposefully generic and will provide fundamental information that can be built upon in other chapters of this book. I will begin by discussing the fibre–matrix interface, after which I will also briefly discuss delamination in composite materials. This will be followed by a section on strain energy release, which is important groundwork in assessing and designing composites to be fracture tough. Finally, to contextualise the importance of interfaces in composite materials, I will conclude the chapter with examples of composite failure and the role of the interface in each.

1.2 The fibre–matrix interface

The interface plays a vital role in ensuring the continuity of components in composite materials and structures. The more well bonded an interface, the more effective the transfer of mechanical energy between composite components. A perfect interface is one that maximises material compatibility, remaining intact under loading such that load-induced failure occurs within the component materials, rather than at the interface. The ideal interface therefore preserves the damage tolerance of the weakest material, and if well enough preserved during failure, has the ability to divert or even to retard crack propagation.

As discussed by Pippel and Woltersdorf [1], in fibre-reinforced plastics (FRPs), interfacial shear stress, τ, controls (to an extent) energy dissipation between the fibre and matrix components, as transmitted by friction. The effect of τ is more easily comprehended using a simplified stress–strain scheme of a unidirectional FRP under tensile loading, figure 1.1, where stress is indicated by σ and strain by ε.

The initial part of the stress–strain curve in figure 1.1 shows linear elastic deformation, where no irrecoverable damage occurs within any of the composite components or at the interface. The first divergence from linearity occurs at the critical stress, σ_{crit}, where matrix cracking commences. This is followed by a further

doi:10.1088/978-0-7503-5688-6ch1

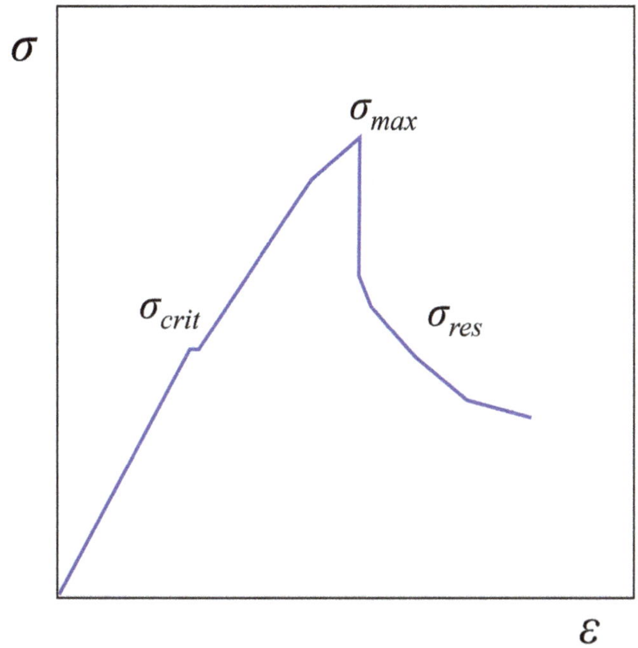

Figure 1.1. Example stress–strain scheme of a unidirectional fibre-reinforced plastic under tensile load.

period of proportionality between loading and material deformation. This is a consequence of the intact continuous unidirectional fibres bridging any matrix cracks and retarding any further development of stress intensity at the crack tip. This critical stress, σ_{crit}, can be calculated according to the energy balance shown in equation (1.1). In this equation, Γ_m is the surface energy of the matrix, V_f is the fibre volume fraction, V_m is the matrix volume fraction, E_f is the fibre Young's modulus, E_m is the matrix Young's modulus, E is the composite Young's modulus, and r_f is the fibre radius.

$$\sigma_{crit} = \left[\frac{6\tau\Gamma_m V_f^2 E_f E^2}{V_m E_m^2 r_f} \right]^{\frac{1}{3}} \tag{1.1}$$

The maximum stress, σ_{max}, can be computed using Weibull parameters (σ_0, l_0 and the Weibull modulus m) in accordance with equation (1.2), assuming that shear stress transmitted by the continuous fibres is both already known and constant.

$$\sigma_{max} = V_f \left(\frac{2\sigma m \tau l_0}{r_f(m+2)} \right)^{\frac{1}{m+1}} \left(\frac{m+1}{m+2} \right) \tag{1.2}$$

Following the maximum stress, there can be a sharp drop in stress. If this drop in stress is not significant, there can be sliding resistance, which occurs as fibres pull out from the composite. This signifies a stress residue, σ_{res}, computed in accordance with equation (1.3), where $\bar{h}$ is an estimated fibre pull-out length.

$$\sigma_{res} = \frac{2V_f\tau\bar{h}}{r_f} \tag{1.3}$$

Equations (1.1) to (1.3) assume that there is no binding across the interface equalling τ. A low value of σ_{res} indicates that the interfaces of fibres are no longer ideal (i.e. damaged or debonded) while a high value of σ_{res} indicates that even though there has been a cross-fracture, the fibres within the matrix are still well bonded and as such require additional energy to be pulled out from the matrix on continued loading.

1.3 Composite delamination

Fibre-reinforced plastic laminates may experience delamination as a load-bearing failure mode, or through exposure to environmental factors including heat, moisture ingress (causing swelling), and freeze-thaw cycles. Thermally induced delamination is often a consequence of poor laminate compatibility, arising through the use of laminating materials with dissimilar coefficients of thermal expansion. Delamination problems may occur during curing, but can also be observed in post-cure thermal events such as those previously mentioned. Delamination is often first noticed through visible cracking, laminate peeling, edge flaking, blistering and bubbling, and in some cases, discolouration. These may instigate from areas where there has been poor surface preparation (e.g. contaminants or residual moisture), resulting in poor bonding and losses in adhesion over time and use.

Mechanical stresses from load-bearing applications as well as direct impaction events give rise to matrix cracks, bending cracks, and shear cracks. These in turn can induce delamination failure, which is fundamentally a result of high interlaminar stresses coupled with low through thickness strength, σ_{33}. Since in laminated FRPs, the fibres are rarely arranged to provide reinforcement in the through thickness direction, the laminates hold themselves together via a weak matrix material. In addition, common matrix materials such as cross-linked epoxies are relatively brittle and thus delamination can occur at high rates, leading to catastrophic structural failure. Put simply, mechanical delamination results fundamentally from:

- the imposition of through (out-of-plane) thickness loading,
- stresses developed due to geometry, and
- stress concentrators owing to laminate connections and discontinuities.

1.3.1 Delamination due to through thickness loading

As mentioned, laminates are weak in the thickness direction (out-of-plane direction) as typically there are no fibres reinforcing the laminated composite through the thickness. As such, through thickness resistance to load is limited by the strength of interlaminar adhesion (i.e. the strength of the matrix polymer). Wisnom [2] discusses two predominating cases for delamination, the first being due to through thickness loading, which may occur through an adhered load-bearing structure as shown in figure 1.2. In this figure two examples are provided, the first where load is applied to an L-piece adhered to the surface of the laminate (Figure 1.2(a)) and the second

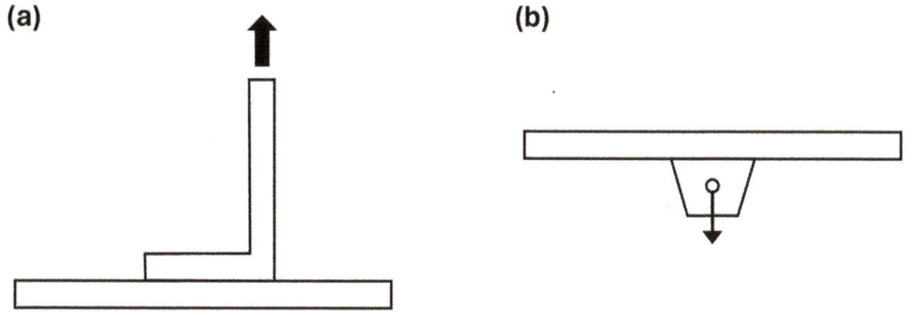

Figure 1.2. Examples of through thickness loading from (a) load imposed on a single L-piece adhered to the surface of the laminate and (b) loading imposed on a lug adhered to the surface of the laminate.

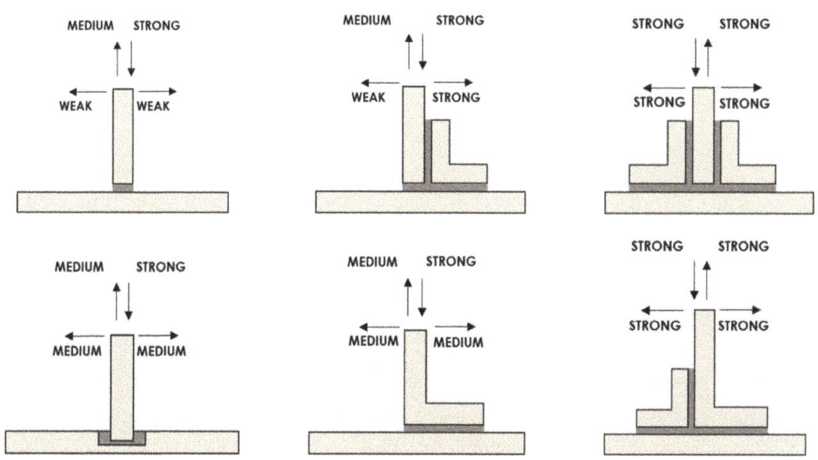

Figure 1.3. Examples of T-joins and their performance expectations with respect to loading direction. Reproduced from [3].

where load is borne by a lug adhered to the surface of the laminate (figure 1.2(b)). In each case, load is transferred to the laminate surface, which develops out-of-plane stresses within the laminate, leading to delamination.

There are several joining mechanisms that develop out-of-plane stresses. Examples include T-joins, some of which are shown in figure 1.3, which also shows their relative performance expectations [3]. Others may include corner joins, as shown in figure 1.4, which also shows their relative performance expectations [3]. Free edge effects can be especially problematic with corner joins, since out-of-plane loading at a free edge is likely to suffer delamination [4] more so than out-of-plane loading away from the free edge (e.g. using T-joins). Weaknesses at free edges are influenced by factors such as cutting method [5], which can to an extent be mitigated through the application of adhesives to cut free edges [6]. Untreated free edges are in addition more prone to delamination through moisture ingress and thermal expansion [7].

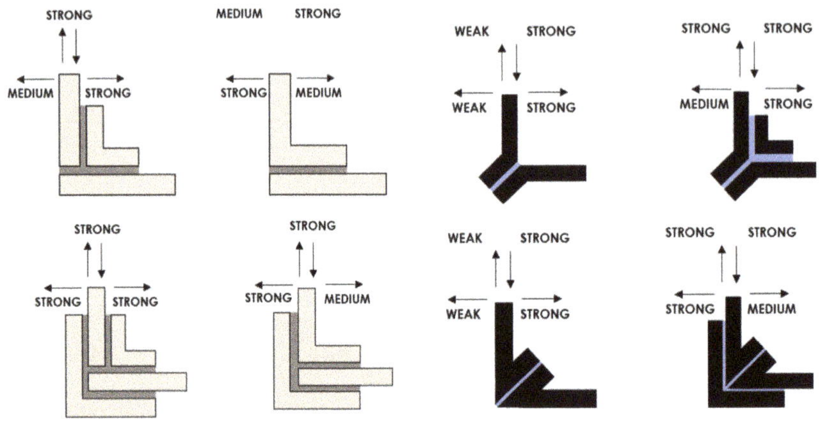

Figure 1.4. Examples of corner joins and their performance expectations with respect to loading direction. Reproduced from [3].

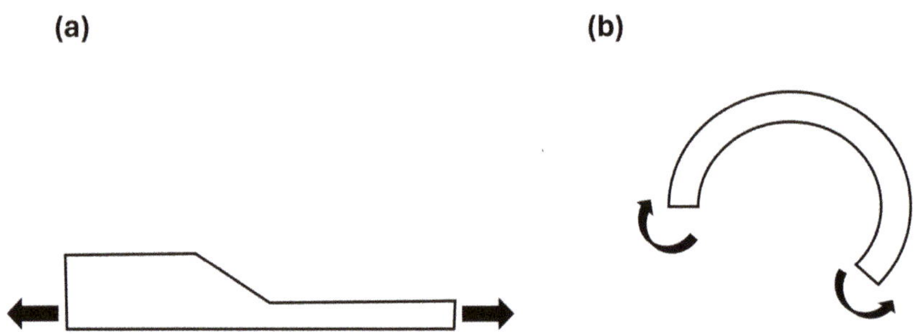

Figure 1.5. Examples of geometrically imposed loading from (a) tapered laminate and (b) a curved laminate subjected to bending moments against its natural curvature.

1.3.2 Delamination due to stresses from geometry

Wisnom [2] further discusses how the geometry of a laminated structure can lead to stress concentrations within laminates, leading to delamination failure. In the first case shown in figure 1.5(a), tapered laminates in tension develop interlaminar stresses as a result of asymmetrical deformation in the region of the taper, resulting in delamination failure. In the second case shown in figure 1.5(b), bending forces acting against the natural curvature of a curved laminate will induce shearing forces between the laminates, again leading to delamination failure. While direct load bearing will have an obvious effect on the development of bending moments, moments can also be developed through swelling and expansion due to water ingress and heat. Each of these in effect can give rise to the evolution of interlaminar stresses, increasing the likelihood of delamination.

1.3.3 Delamination due to laminate connections and discontinuities

Laminated composites are often joined to enable the engineering of larger-scale structures. Joining methods include mechanical fixings such as bolts as well as

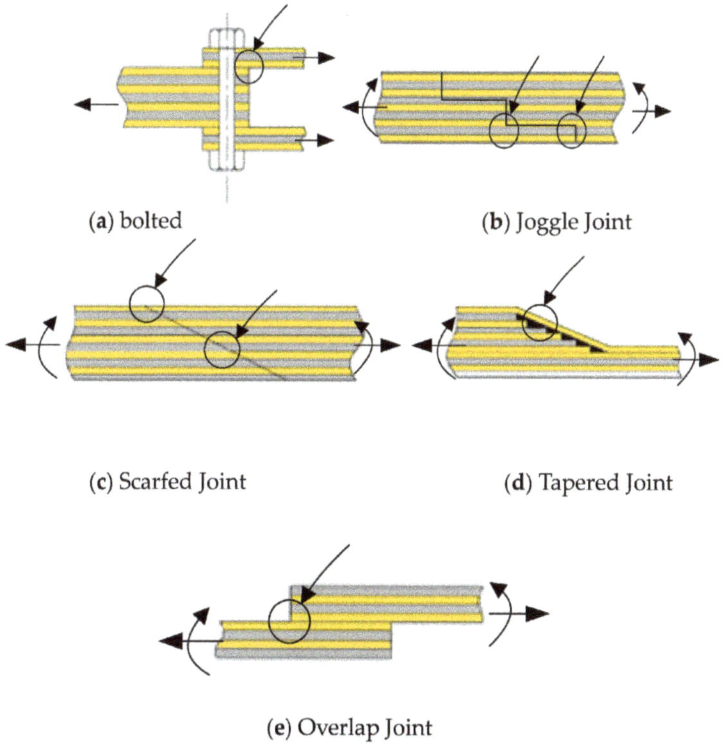

Figure 1.6. Examples of joining methods in composites showing (a) bolted (mechanical) connecters, (b) joggle joints, (c) scarfed joints, (d) tapered joints, and (e) overlap joints. Areas of stress concentration are indicated by circles and arrows. Reproduced from [8]. CC BY 4.0.

composite joining (e.g. lap joints, joggle joints, scarf joints, and tapered ply-drops; figure 1.6). These sites of joining, being discontinuities, are stress concentrators [8] and can be the nucleation points for delamination.

1.3.4 Internal delamination: fundamental theory

Fracture mechanics approaches can be used to develop delamination theory since internal delamination (including edge delamination) is in essence very similar to crack propagation. As such, stable crack growth formulations have been developed in terms of energy release rates and stress intensity factors. The fracture toughness of composites can in turn be characterised in terms of the fracture energy per unit of new surface that is created on delamination. The conventional fracture mechanics energy release rates G_{I}, G_{II}, and G_{III} refer to failure modes I, II, and III, those being opening, shearing (sliding), and anti-plane shear (tearing) modes, respectively. Each failure mode is illustrated in figure 1.7. The corresponding critical energy release rates can then be, following convention, expressed as G_{IC}, G_{IIC}, and G_{IIIC}, respectively.

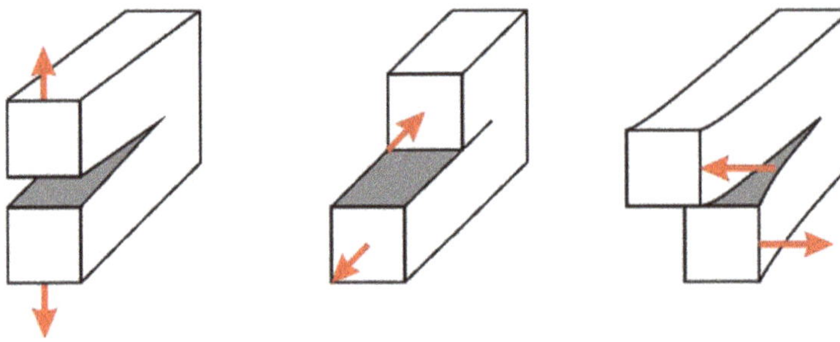

Figure 1.7. Illustrations of solid material fracture modes I (crack opening), II (shearing or sliding) and III (tearing, or anti-plane shearing). Reproduced from [9]. CC BY 4.0.

Bolotin [10] discusses an analytical approach to the theory for the cases of stable internal delamination between two laminates with similar properties. Delamination was considered as being oriented to the principle axes and while the strain energy release rates of elastic materials should ordinarily be additive, equation (1.4), Bolotin describes this as being devoid of true meaning since laminated composites typically exhibit high levels of anisotropy. Here, G_c is the critical total strain energy release rate.

$$G_c = G_{\mathrm{I}} + G_{\mathrm{II}} + G_{\mathrm{III}} \tag{1.4}$$

Since the work of fracture is dependent on the specific fracture mode experienced, equation (1.5) is recognised as often being a successful fit with experimental data [8, 10, 11]. In this equation, m_{I}, m_{II}, and m_{III} are empirical exponents.

$$\left(\frac{G_{\mathrm{I}}}{G_{\mathrm{IC}}}\right)^{m_{\mathrm{I}}} + \left(\frac{G_{\mathrm{II}}}{G_{\mathrm{IIC}}}\right)^{m_{\mathrm{II}}} + \left(\frac{G_{\mathrm{III}}}{G_{\mathrm{IIIC}}}\right)^{m_{\mathrm{III}}} = 1 \tag{1.5}$$

While the relationship in equation (1.5) is in essence, phenomenological [8], it assumes that delamination growth is in-plane normal to the crack front, and that the ratio between G_{I}, G_{II}, and G_{III} is invariable. To effectively use equation (1.5), individual energy release rate components (i.e. individually as modes I, II, and III) need to first be determined. Further information on the strain energy release rate, G, is provided in the following section (section 1.4).

1.4 Strain energy release rate

A strain energy release rate, G, is defined as the instantaneous release of strain energy, U, per unit area, A, of a cracked material, equation (1.6). The crack area can also be reduced to a one-dimensional crack length, a, in two-dimensional problems, keeping the width of the sample constant as the second dimension of area, b. When the strain energy release rate reaches a critical value, G_c, the crack will grow provided the available energy release rate is equal to, or above, G_c, equation (1.7).

$$G = -\frac{\partial U}{\partial A} \tag{1.6}$$

$$G \geqslant G_c = -\frac{1}{b}\frac{\partial U}{\partial a} \tag{1.7}$$

G_c is thus also termed the fracture energy, and is a material property that is proportional to the failure deformation that occurs at a crack tip. This may be either a brittle or a plastic failure deformation. If brittle, then the Griffith model for brittle failure can be used, where the work, γ_s, needed to separate each of the two surfaces is equal to G_c, equation (1.8). The Griffith model assumes a perfectly elastic solid with no plasticity in the failure mode. If there is, however, an element of plastic failure, then the Irwin model, equation (1.9), is more typically valid as it includes the plastic work, γ_p, needed to develop unstable crack growth at a plastic crack front.

$$G_c = 2\gamma_s \text{ (brittle fracture)} \tag{1.8}$$

$$G_c = 2\gamma_s + \gamma_p \text{ (coupled brittle–plastic fracture)} \tag{1.9}$$

The strain energy release rate is directly related to the stress intensity factor, K, of a material, equation (1.10), where σ is the applied stress. To differentiate K between modes I, II, and III fracture K will be expressed as K_{I}, K_{II}, and K_{III}, respectively. Similarly, the strain energy release rate, G, is typically expressed as G_{I}, G_{II}, and G_{III}, for cases of modes I, II, and III fracture, respectively. The relationships between G_{I} and K_{I}, G_{II} and K_{II}, and G_{III} and K_{III} are provided in equations (1.11), 1.12, and (1.13), respectively, where E is the Young's modulus, ν is the Poisson's ratio, and μ is the shear modulus.

$$K = \sigma\sqrt{\pi a} \tag{1.10}$$

$$G_{\text{I}} = \frac{K_{\text{I}}^2}{E} \text{ (plane stress) ... or, } G_{\text{I}} = \frac{K_{\text{I}}^2(1-\nu^2)}{E} \text{ (plane strain)} \tag{1.11}$$

$$G_{\text{II}} = \frac{K_{\text{II}}^2}{E} \text{ (plane stress) ... or, } G_{\text{II}} = \frac{K_{\text{II}}^2(1-\nu^2)}{E} \text{ (plane strain)} \tag{1.12}$$

$$G_{\text{III}} = \frac{K_{\text{III}}^2}{2\mu} \tag{1.13}$$

For fracture to propagate, the strain energy release rate, G, must be equal to or above its critical value, G_c, such that $G_{\text{I}} \geqslant G_{\text{Ic}}$, $G_{\text{II}} \geqslant G_{\text{IIc}}$, and $G_{\text{III}} \geqslant G_{\text{IIIc}}$, each of which can be related to the critical stress intensity factor according to equations, (1.14), (1.15), and (1.16), respectively.

$$G_{\text{Ic}} = \frac{K_{\text{Ic}}^2}{E} \text{ (plane stress) ... or, } G_{\text{Ic}} = \frac{K_{\text{Ic}}^2(1-\nu^2)}{E} \text{ (plane strain)} \tag{1.14}$$

$$G_{\text{IIc}} = \frac{K_{\text{IIc}}^2}{E} \text{ (plane stress) ... or, } G_{\text{IIc}} = \frac{K_{\text{IIc}}^2(1-\nu^2)}{E} \text{ (plane strain)} \quad (1.15)$$

$$G_{\text{IIIc}} = \frac{K_{\text{IIIc}}^2}{2\mu} \quad (1.16)$$

Of these, modes I and II fracture are more commonly researched by composites engineers, primarily because it can be very difficult to achieve mode III fracture under laboratory testing conditions. There are several methods that have been used to determine both mode I and mode II fracture in composite materials, which are summarised in table 1.1. Several of these methods are covered in greater detail in chapter 7 of this book.

Perhaps the most common method to date for mode I testing is the double cantilever beam (DCB) method (see chapter 7 for more details). The general energy method, shown in equation (1.17), can be used for G_{Ic} computation, but derived compliance methods may also be used. These are shown in equations (1.18) (compliance calibration method) and (1.19) (modified compliance calibration method), where P is the load, δ is the load point displacement, C is the compliance of the DCB specimen calculated as $\frac{\delta}{P}$, n is the slope of $\log C$ against $\log a$, h is the specimen thickness, and A_1 is the slope from a plot of $\frac{a}{b}$ against $C^{\frac{2}{3}}$. An alternative method is a modified beam method, shown in equation (1.20).

$$G_{\text{I}} \geqslant G_{\text{Ic}} = -\frac{1}{b}\frac{\partial U}{\partial a} \quad (1.17)$$

$$G_{\text{Ic}} = \frac{nP\delta}{2ba} \quad (1.18)$$

$$G_{\text{Ic}} = \frac{3P^2C^{\frac{2}{3}}}{2A_1bh} \quad (1.19)$$

$$G_{\text{Ic}} = \frac{3P\delta}{2ba} \quad (1.20)$$

In terms of mode II testing, the end-notched beam (ENF) test is one of the more commonly used for composites (see chapter 7 for a more detailed overview). Similar to mode I (DCB) testing, two generic methodological bases are used to determine the critical strain energy release rate in mode II. One set of methods is compliance based while another set of methods is based on beam theory. The compliance calibration method is often used to compute G_{IIc} in ENF samples. This is shown in equation (1.21), where P is load, b is coupon width, a is crack length, and the constant m is related to compliance, C, such that $m = \frac{C-D}{a^3}$ (noting that D is a constant). Direct beam theory methods are shown in equation (1.22), where δ is displacement and L is the coupon length. Finally, corrected beam theory methods can be used, equation

Table 1.1. Common methods for the determination of mode I and mode II fracture properties of composite materials.

Test method described in:	Source
MODE I	
Standard Test Method for Mode I Interlaminar Fracture Toughness of Unidirectional Fiber-Reinforced Polymer Matrix Composites	ASTM D5528-13 [12]
Fibre-reinforced plastic composites—Determination of mode I interlaminar fracture toughness, GIC, for unidirectionally reinforced materials	ISO 15 024:2023 [13]
Adhesives—Determination of the mode 1 adhesive fracture energy of structural adhesive joints using double cantilever beam and tapered double cantilever beam specimens	ISO 25 217:2009 [14]
Adhesives—Determination of dynamic resistance to cleavage of high-strength adhesive bonds under impact wedge conditions—Wedge impact method	ISO 11 343:2019 [15]
MODE II	
Standard Test Method for Determination of the Mode II Interlaminar Fracture Toughness of Unidirectional Fiber-Reinforced Polymer Matrix Composites	ASTM D7905/ D7905M-19e1 [16]
New test method for determination of the mode II interlaminar fracture toughness of unidirectional fiber-reinforced polymer matrix composites using the end-notched flexure (ENF) test	WK22949 A [17]
Fibre-reinforced plastic composites—Determination of the mode II fracture resistance for unidirectionally reinforced materials using the calibrated end-loaded split (C-ELS) test and an effective crack length approach	ISO 15 114:2014 [18]
Comparison of Delamination Characterization for IM7/8552 Composite Woven and Tape Laminates	Paris *et al* (2003) [19]

(1.23), where the axial composite modulus is calculated as $E_f = \frac{L^3}{4bh^3C_0}$. Here, C_0 is the initial compliance at $a = a_0$.

$$G_{\mathrm{IIc}} = \frac{3P^2\, ma^2}{2b} \tag{1.21}$$

$$G_{\mathrm{IIc}} = \frac{9a^2P\delta}{2b(2L^3 + 3a^3)} \tag{1.22}$$

$$G_{\mathrm{IIc}} = \frac{9a^2P^2}{16b^2h^3E_f} \tag{1.23}$$

Engineered composites may experience mixed modes of failure. Mapping the magnitudes of strain energy released during a mode I or II failure against one another can be a useful practice in composites design as it provides information on the relative fracture strengths of different composite types and formulations. Figures 1.8 and 1.9, for example, map the experimental values for G_I against those for G_{II} for a range of carbon FRP (CFRP) and glass FRP (GFRP) composites, respectively, using data from [20–34]. In these figures, CF refers to carbon fibre, GF refers to glass fibre, UD refers to a unidirectional fibre orientation, PA6 is polyamide 6 (matrix), PPS is polyphenylene sulphide (matrix), and PE is polyester (matrix). It is evident from figure 1.8 that fibre orientation affects the distribution of G_{I} and G_{II} properties in CFRP composites (CF(biaxial)/epoxy against CF(UD)/epoxy), as does

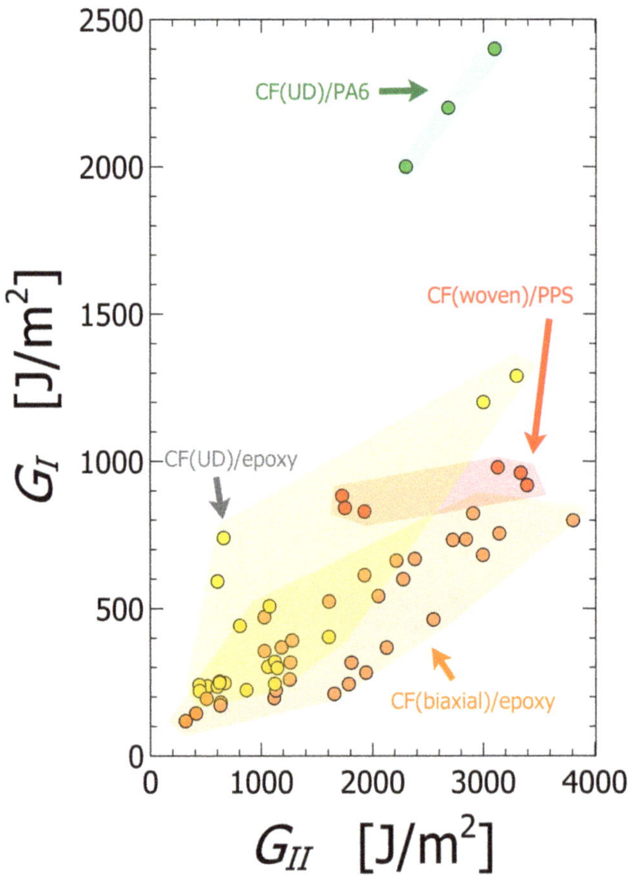

Figure 1.8. G_{I} plotted against G_{II} for carbon fibre-reinforced plastic (CFRP) composites comprising different fibre/matrix formulations using data published in [20–27]. Here PA6 is polyamide 6 and PPS is polyphenylene sulphide.

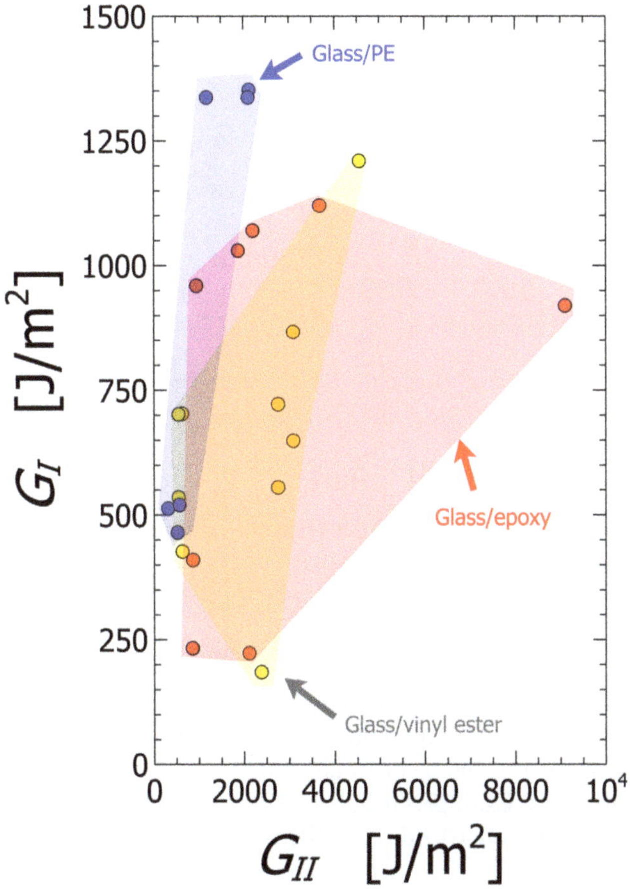

Figure 1.9. G_{I} plotted against G_{II} for glass fibre-reinforced plastic (GFRP) composites comprising different fibre/matrix formulations using data published in [28–34]. Here PE is polyester.

the type of matrix material used, with PA6 matrix composites yielding significantly higher G_{I} to G_{II} values than epoxy matrix composites. This is also noticeable in figure 1.9 where larger relative differences in G_{I} to G_{II} can be noted in PE matrix composites as compared against epoxy matrix composites. Fibre orientation and matrix selection should thus be considered within the design process when the relative modes of failure are of specific importance.

In general, G_{I} will be lower than G_{II} in most engineering CFRP and GFRP composites as shown in the scatterplot of G_{I} against G_{II} for general carbon and glass fibre-reinforced polymer composites (figure 1.10). This is an expected outcome since the peel resistance (mode I fracture) of well-adhered materials tends to be lower than the shear resistance (mode II fracture). This in turn is because stresses in a peel-type test (e.g. DCB) are concentrated over a very small area of the total bonded area, whereas in shear-type tests (e.g. ENF) a greater proportion of the bonded area is involved in bearing load. Finally, both G_{I} and G_{II} often tend to be linearly correlated, albeit with significant levels of scatter.

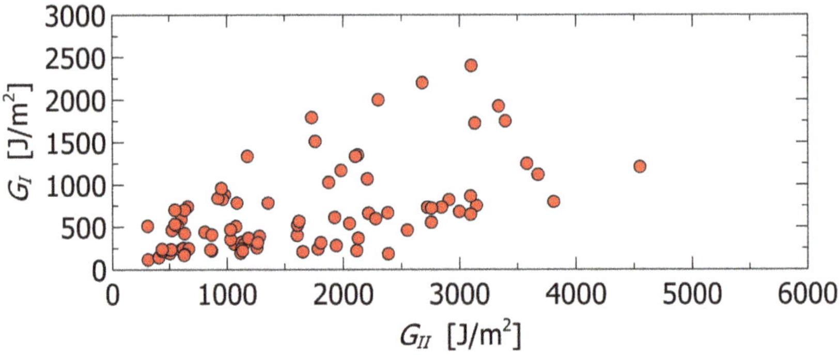

Figure 1.10. Scatterplot showing G_{I} against G_{II} for general carbon and glass fibre-reinforced polymer composites.

1.5 Examples of interfacial failure

It may be useful to end this chapter with examples of interfacial failure in composite constructions. Two example areas are provided here; the first will consider helicopter rotor blades and the second composite rocket constructions.

1.5.1 Helicopter rotor blades

Composites have been important in the development of durable helicopter rotor blades, with earlier blades being manufactured as composites comprising wooden core [35] typically from spruce, birch, or douglas fir trees, sheathed by thin form-fitting skins, sometimes fabrics. Examples include the Bell 47 (Bell Aircraft Corporation), the Bensen B8M autogyros and B19 designs (Bensen Aircraft Corporation), and the C8 autogyros (Cierva Autogiro Company). The woods used in these early helicopter blades have good overall properties of stiffness and fatigue, and are low-density materials. Nevertheless, since wood is both hygroscopic and hydrophilic, these blades were prone to swelling and warping, which in turn were enablers for both ply delamination and altered aerofoil tolerances [36]. Moisture diffusion into wood *somewhat* follows Fick's second law of diffusion [37]. The one-dimensional unsteady state of moisture diffusion in a direction perpendicular to the wood surface is described by the partial differential equation, equation (1.24), where C is the water concentration, t is time, x is the direction of flux, and D is the diffusivity. As can be seen in the equation, the accumulation $\frac{\partial C}{\partial t}$ is proportional to D and the second derivative of C (i.e. $\frac{\partial^2 C}{\partial x^2}$).

$$\frac{\partial C}{\partial t} = \frac{\partial}{\partial x}\left(D\frac{\partial C}{\partial x}\right) \tag{1.24}$$

Blades making use of alloys at that time were seen as solving some of the problems of wooden blades. As such, composite blades comprising skins (either metal or fibreglass) coupled to aluminium alloy or Nomex honeycomb gained greater popularity in the 1970s. The Sea King and Westland (Westland Helicopters Ltd) blades, for example, comprised alloy skins connected to Nomex honeycomb cores, while early Lynx (Westland Helicopters Ltd) blades were made up from titanium spars coupled to glass-fibre skins [36]. But there are specific identifiable problems in terms of failure and fracture in using metals: (a) metal alloy composite blades have a low fatigue life and as such require routine replacing typically within a service life of 4–6000 h (though certain models provide guidance for metal blade replacement every 2200 service hours, e.g. the Robinson R44 (Robinson Helicopter Company Inc.) [38]) and (b) when a fatigue crack forms, complete blade failure usually occurs quickly and catastrophically [39]. One example of Robinson R44 blade failure can be cited from an incident that took place on 10 July 2019 on a flight between Lac de la Bidiere, Quebec, to Sainte-Sophie, Quebec, Canada. The blades of this helicopter comprised an aluminium alloy honeycomb core bordered by a stainless steel spar and stainless steel skins on both the upper and lower blade faces. The perceived cause of blade failure was debonding at the interface between the skin and the spar, leading to further debonding between the skin and the honeycomb core. This problem occurred prior to the 2200 service hours or 12-year recommendation by the Robinson Helicopter Company R44 manual and was potentially a reason for flux instabilities, eventuating in the R44 crash and the death of all passengers aboard the helicopter [40].

Fibre-reinforced plastic composites began to replace metal alloy-based blades in the 1970s to 1980s as they offered improvements in fatigue life [41], and additionally tended to avoid catastrophic failure unlike their metal alloy blade counterparts. A generic example of an FRP composite helicopter blade section is shown in figure 1.11 that comprises a foam core, a CFRP composite central stiffener, FRP skin covering the core, a GFRP composite thick section at the leading edge, and additional thin-sheet stainless steel leading edge protection. FRP composites are high-performance materials with lower densities than many metal alloys. As such there is also current interest in the 'composite light-weighting' of helicopters and planes, enabling reduced fuel use and potentialising increased flight time. In fact, the rotor blades used on the Bell 206B JetRanger as supplied by Van Horn Aviation are reported to have an extended service life of 18 000 h, which is more than triple that of standard metal alloy-based blades [42]. Standard FRP composite blades (employing aerofoils SC1095 and SC1094 R8) used in UH-60 Black Hawk helicopters, for example, are

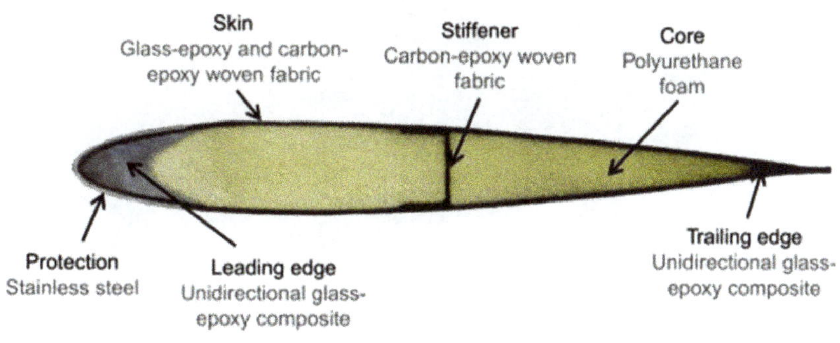

Figure 1.11. Example cross-section of a generic FRP composite helicopter blade. Reprinted from [43]. Copyright (2017), with permission from Elsevier.

comprised of a titanium spar with a GFRP outer contour. Wide chord blades (using aerofoils SC2110 and SSCA09) have an all-composite graphite/glass tubular spar [46, 47]. While these blades are designed well for rotary motion, as demonstrated on 22 February 2022, the UH-60 Blackhawk blades are unable to take rolling loads in air. As reported in [48] in an incident where a UH-60 Blackhawk rolled to its side the main rotor blades broke apart, which in turn caused impact damage to a nearby (lead) aircraft with direct damage to the main and tail rotor blades. While it can be inferred from this example that sharp variations in airflow can cause irreparable damage to composite helicopter blades, direct impaction events are also regularly recorded in flight fail literature. An example shown in figure 1.12 is from an incident involving two helicopters, an AS 332 Super Puma, which collided with a Eurocopter EC 155 that was stagnant on the ground. As can be seen in the figure, the blades of the stagnant EC 155 splayed irreparably from dynamic impact with the incoming AS 332.

Interfacial failure between FRP composite laminates can also occur through either hard (stones, bullets, etc.) or soft (birds, hailstones, etc.) impact events [43]. These impacts can catastrophically damage the structural integrity of the composite skins causing failure at interfaces including internal delamination of the skins, delamination failure at the skin-core interface, fibre failure, and matrix cracking. An example from an experimental study by [45] is shown in figure 1.13, where the penetrating bullet caused skin delamination, fibre breakage, core collapse, and delamination failure at the skin-core interface.

The stiffness and fatigue [41] performance offered by advanced composites such as continuous fibre carbon- and glass-reinforced plastics are a reason more modern helicopters such as the Airbus H160, the Bell Booeing V-22, and the Airbus EC145/BK117 are manufactured from a significant fraction of composite. The Airbus H160, for example, is the first helicopter made almost entirely from composite

Figure 1.12. Damage to the stagnant helicopter after direct dynamic impact with another helicopter. The red line indicates the direction of impact. Reproduced from [44], with permission from Springer.

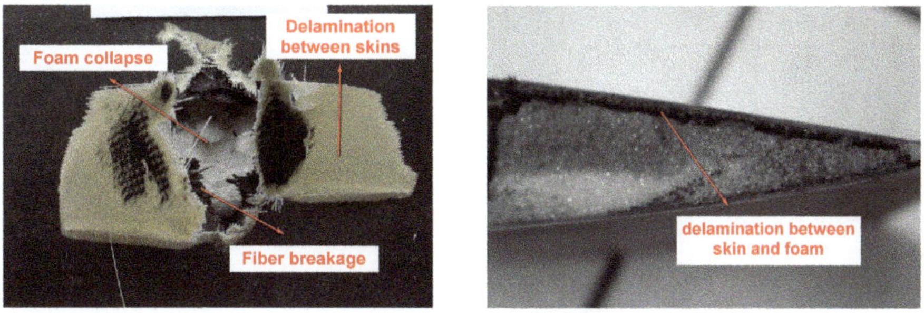

Figure 1.13. Experimental ballistic test on an FRP composite helicopter blade providing examples of skin delamination, fibre breakage, core collapse, and delamination failure at the skin-core interface. Reproduced from [45]. CC BY 4.0.

material, while the Bell Boeing V-22 includes composite rotor blades and a composite fuselage stiffened skin [49]. Additional factors including high thermal stability, thermal conductivity, and low mass area make composites an attractive material for future helicopter part designs.

1.5.2 Rockets

Materials selected for rockets will typically need to meet certain functional requirements including high specific stiffness and strength, impact damage tolerance, thermal stability, ease of manufacture, retained functionality under the expected service conditions, cost, etc. (figure 1.14). Materials meeting the requirements of thermal protection can include carbon/carbon matrix composites, silica tiles, and polyurethane foams. Silicon carbide matrix composites reinforced by silicon carbide fibre (SiCf/SiC composite) and carbon fibre/SiC composites are good candidate thermostructural materials. Good candidate structural alloys include titanium alloys, aluminium alloys, magnesium/magnesium–lithium alloys, age-hardened iron–nickel–steel alloys (such as Maraging 250), and low-carbon alloyed steels with high toughness and yield strength and that weld well (e.g. 15CDV6 as alloy steel) [50]. Rocket motor casings, canards, wings, nose cones, and tail fins can use a variety of composites including carbon fibre-reinforced epoxy and Kevlar fibre-reinforced epoxy, and in areas of required impact damage tolerance, GFRPs.

Selected materials are typically those that are sufficiently able to exceed the design requirement. For example, a rocket motor casing chamber may be designed to have a chamber pressure (P_c), equation (1.25), which is lower than the design pressure (P_d) that will be used for design calculations. P_d will typically include a factor of safety, and will always be lower than the maximum allowable chamber pressure. The internal chamber pressure leads to loading in the axial and hoop directions of the chamber, which can be crudely approximated using equations (1.26) and (1.27), respectively. In these equations, F is the engine thrust, A_t is the hoop area, C_f is the thrust coefficient, F_a is the axial force per unit length under the design pressure P_d, F_h is the hoop force per unit length under the design pressure P_d, and D is the mean diameter of the rocket motor casing chamber.

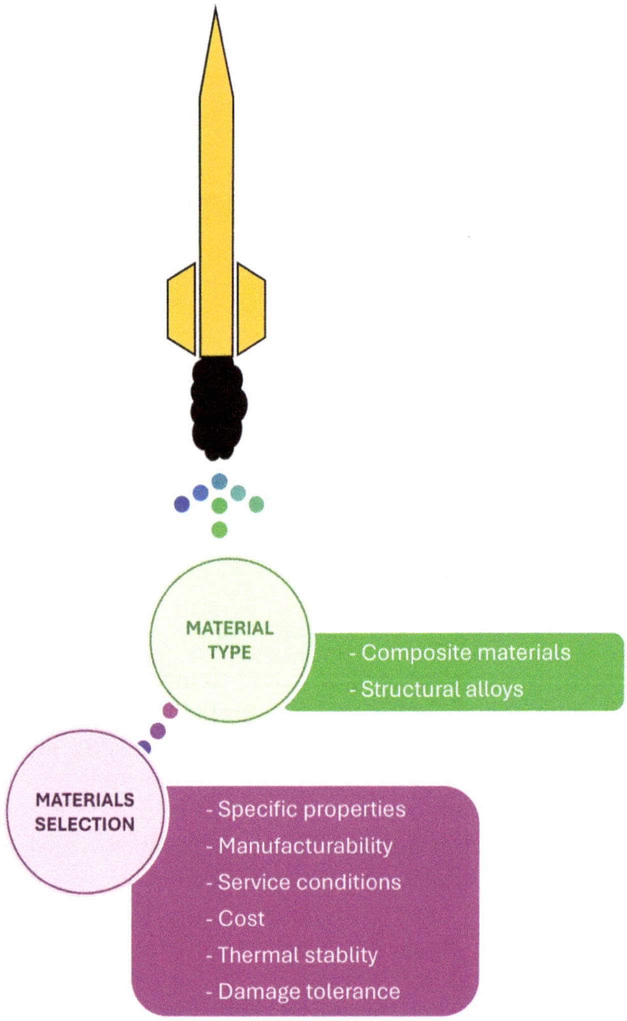

Figure 1.14. Materials selection considerations and types of materials in the design of rockets.

$$P_c = \frac{F}{A_t C_f} \tag{1.25}$$

$$F_a = \frac{P_d D}{4} \tag{1.26}$$

$$F_h = \frac{P_d D}{2} \tag{1.27}$$

In equation (1.25), the thrust, F, is calculated as a product of the mass flow rate, $\dot{m}$, and the exit velocity, V_e, added to the product of the exit pressure, p_e, less the free stream pressure, p_0, and the exit area, A_e, equation (1.28). Here, $\dot{m}$ is calculated according to equation (1.29), where A^* is the minimal throat area of the casing, T_t is the total temperature, γ is the specific heat ratio, R is the gas constant, and p_t is the total pressure. The exit velocity is calculated according to equation (1.30), where T_e is the exit temperature and M_e is the Mach number on exit.

$$F = \dot{m}V_e + (p_e - p_0)A_e \tag{1.28}$$

$$\dot{m} = \frac{A^* p_t}{\sqrt{T_t}}\sqrt{\frac{\gamma}{R}}\left(\frac{\gamma+1}{2}\right)^{-\frac{\gamma+1}{2(\gamma-1)}} \tag{1.29}$$

$$V_e = M_e\sqrt{\gamma R T_e} \tag{1.30}$$

Rocket temperatures can be very high and as heat is transmitted to the composite thermal stresses may develop. Following composite laminate theory [3], these stresses can be calculated in the kth lamina in accordance with equation (1.31), when the temperature change, ΔT, is known. Here, thermal stresses in x, y, and xy are denoted σ_x^T, σ_y^T, and σ_{xy}^T, respectively; the matrix $\left[\bar{Q}\right]$ is the reduced transformation matrix of the kth lamina; midplane strains are denoted ε_x^T, ε_y^T, and ε_{xy}^T, in x, y, and xy, respectively; the midplane curvatures are recognised as K_x, K_y, and K_{xy}, in x, y, and xy, respectively; and α_x, α_y, and α_{xy} are the coefficients of thermal expansion in x, y, and xy, respectively.

$$\begin{bmatrix}\sigma_x^T\\ \sigma_y^T\\ \sigma_{xy}^T\end{bmatrix}_k = \left[\bar{Q}\right]_k\begin{bmatrix}\varepsilon_x^0\\ \varepsilon_y^0\\ \varepsilon_{xy}^0\end{bmatrix} + z\left[\bar{Q}\right]_k\begin{bmatrix}K_x\\ K_y\\ K_{xy}\end{bmatrix} - \left[\bar{Q}\right]_k\begin{bmatrix}\alpha_x\\ \alpha_y\\ \alpha_{xy}\end{bmatrix}_k \Delta T \tag{1.31}$$

Designing rocket parts using composites requires a consideration of the effects of thermal stresses not only on the properties of the material as a function of extreme heating or extreme cooling, but also in relation to coupled effects that may result in interface failure and delamination such as impaction events [51]. While the equations to calculate thermal stress are best used following composite laminate theory, crude approximations (to help gauge an appropriate design space) can also be made as $\sigma^T = \frac{\alpha E \Delta T}{2(1-\gamma)}$, where E is an averaged Young's modulus of the composite material [52].

An interesting survey can be made of fracture modes from a rocket crash from The University of Edinburgh's rocketry team, Endeavour. The team designed [53] and manufactured several rockets, including recently Darwin III, figure 1.15(a), a composite rocket that travelled at 2049 kmh^{-1}, reaching an altitude of 6983m at the European Rocketry Challenge held at Ponte de Sor, Portugal in 2022. Due to a problem with the parachute deploying, the rocket impacted the soil nose-first at 903 $\mathrm{km\,h}^{-1}$, figure 1.15(b), resulting in significant damage, figures 1.15(c)–(f). Macro-damage of the composite that is immediately noticeable includes transverse external fracture including visible fibre fracture of the filament wound fuselage, figure 1.15(c), most likely arising due to the high compressive forces through the axis

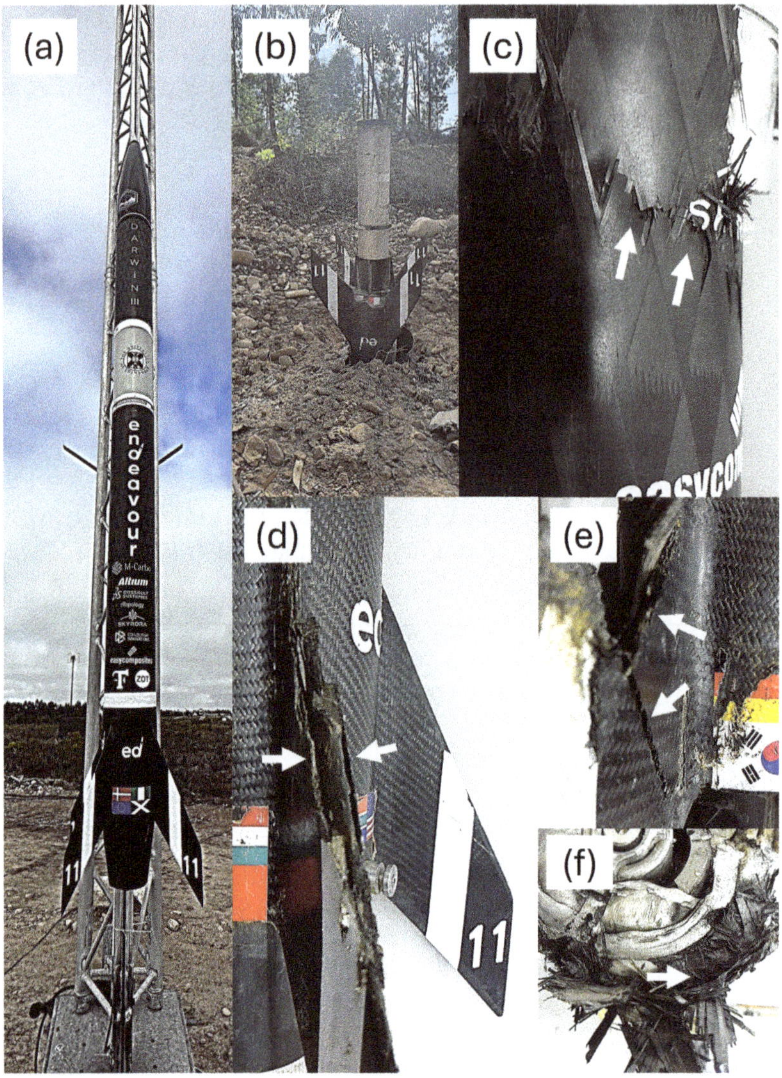

Figure 1.15. (a) Darwin III rocket pre-launch (b) nose-first impaction into the ground following a failed parachute deploy, (c) transverse external fracture including visible fibre fracture of the main body tube, (d) delamination fracture in the fins, (e) coupled delamination-cross fracture in the fins, and (f) composite splaying. Images provided courtesy of Endeavour, The University of Edinburgh's rocketry society.

of the cylinder body due to nose-first impaction with the ground. Fracture in constrained thin-walled structures such as this will often be a result of shear buckling failure of the fibres and the subsequent formation of kink-bands within the laminate [54, 55] (see chapter 9 section 9.5). Delamination can be observed in the fins, figure 1.15(d), while coupled delamination-cross fracture can be observed in figure 1.15(e). Finally, composite splaying can be observed in figure 1.15(f), possibly arising from edge impaction with the ground.

Darwin III is an interesting case of structural failure as it is due to full body impaction. Structural failures in rockets are fairly common. Dynamic stresses, for example, can arise through excessive vibrations and these are known to be highly detrimental to composite interfaces [57]. Moreover, high pressures potentialise burst-related micro- to macroscale damage within the fuselage material [56]. To circumvent problems from pressure, engineers logically use information on the different failure modes from composite burst tests as part of the redesign process [58]. A combination of both GFRP and CFRPs (multireinforced composites) is nevertheless being proposed as a solution to problems with delamination in composite rocket motor casing joints [59]. Martin [60] discusses the reasons for common structural failures in rockets. Ruptures in a filament wound composite (fuselage, nozzle cap, etc.), for example, may be inherited from poor manufacturing considerations such as incorrect filament-to-resin ratios, irregular filament indexing, curing errors, errors in the number of turns of the wound filament, moisture pickup by the filament, over-aged resin formulations, and general poor bonding at component interfaces. Errors at the stages of manufacturing are sometimes overlooked and some have led to catastrophic failures, such as the loosening of a composite overwrapped pressure vessel in the Space X Falcon 9 on 28 June 2015 [61] and the thermomechanical erosion of a carbon/carbon composite nozzle throat insert in Avio's (Vega C) Zefiro 40 rocket motor on 20 December 2022 [62].

References

[1] Pippel E and Woltersdorf J 1996 Interfaces in composite materials Acta *Phys. Pol.* **89** 209–18

[2] Wisnom M R 2012 The role of delamination in failure of fibre-reinforced composites *Phil. Trans. R. Soc.* **370** 1850–70

[3] Alam P 2021 *Composites Engineering: An A-Z Guide* (Bristol: IOP Publishing)

[4] Lagunegrand L, Lorriot T, Harry R, Wargnier H and Quenisset J M 2006 Initiation of free-edge delamination in composite laminates *Compos. Sci. Technol.* **66** 1315–27

[5] Geier N, Xu J, Poor D I, Dege J H and Davim J P 2023 A review on advanced cutting tools and technologies for edge trimming of carbon fibre reinforced polymer (CFRP) composites *Composites B* **266** 111037

[6] Brauning K A, Kunza A, Alarifi I M and Asmatulu R 2020 Mitigations of machine-damaged free-edge effects on fiber-reinforced composites *J. Compos. Mater.* **55** 1621–33

[7] Alam P, Robert C and O Bradaigh C M 2018 Tidal turbine blade composites–a review on the effects of hygrothermal aging on the properties of CFRP *Composites B* **149** 248–59

[8] Huang T and Bobyr M 2023 A review of delamination damage of composite materials *J. Compos. Sci.* **7** 468

[9] Preiβ 2018 Fracture toughness of freestanding metallic thin films studied by bulge testing *Doctoral Thesis* Friedrich-Alexander University FAU University Press, Parallel erschienen als Druckausgabe bei

[10] Bolotin V V 1996 Delaminations in composite structures: its origin, buckling, growth and stability *Composites* **27B** 129–45

[11] Donaldson S U 1985 Fracture toughness testing of graphite (epoxy and graphite) PEEK composites *Composites* **16** 103–12

[12] ASTM D5528-13 2022 *Standard Test Method for Mode I Interlaminar Fracture Toughness of Unidirectional Fiber-Reinforced Polymer Matrix Composites* (West Conshohocken, PA: ASTM International)
[13] ISO 15 024:2023 Fibre-reinforced plastic composites—determination of mode I interlaminar fracture toughness, GIC, for unidirectionally reinforced materials, International Organization for Standardization, Geneva, Switzerland
[14] ISO 25 217:2009 Adhesives—determination of the mode 1 adhesive fracture energy of structural adhesive joints using double cantilever beam and tapered double cantilever beam specimens, International Organization for Standardization, Geneva, Switzerland
[15] ISO 11 343:2019 Adhesives—determination of dynamic resistance to cleavage of high-strength adhesive bonds under impact wedge conditions—wedge impact method, International Organization for Standardization, Geneva, Switzerland
[16] ASTM D7905/D7905M-19e1 2019 *Standard Test Method for Determination of the Mode II Interlaminar Fracture Toughness of Unidirectional Fiber-Reinforced Polymer Matrix Composites* (West Conshohocken, PA: ASTM International)
[17] WK22949 A 2010 *New Test Method for Determination of the mode II Interlaminar Fracture Toughness of Unidirectional Fiber Reinforced Polymer Matrix Composites Using the End-notched Flexure (ENF) Test* (Conshohocken, PA: ASTM International)
[18] ISO 15 114:2014 Fibre-reinforced plastic composites—Determination of the mode II fracture resistance for unidirectionally reinforced materials using the calibrated end-loaded split (C-ELS) test and an effective crack length approach, International Organization for Standardization, Geneva, Switzerland
[19] Paris I, Minguet P J and O'Brien T K 2003 Comparison of delamination characterization for IM7/8552 composite woven and tape laminates *Composite Materials: Testing and Design* (West Conshohocken, PA: ASTM International) Fourteenth Volume, ASTM Stock Number: STP1436
[20] Baere I D, Jacques S, Paepegem W V and Degrieck J 2012 Study of the mode I and mode II interlaminar behaviour of a carbon fabric reinforced thermoplastic *Polym. Test.* **31** 322–32
[21] Reis J P, De Moura M S F S, Moreira R D F and Silva F G A 2019 Pure mode I and II interlaminar fracture characterization of carbon-fibre reinforced polyamide composite *Composites B* **169** 126–32
[22] Ramirez F M G, De Moura M S F S, Moreira D R F and Silva F G A 2021 Experimental and numerical mixed-mode I + II fracture characterization of carbon fibre reinforced polymer laminates using a novel strategy *Compos. Struct.* **263** 113683
[23] Lachaud F, Piquet R and Michel L 1999 Delamination in mode I and II of carbon fibre composite materials: fibre orientation influence *Proc. 12th Int. Conf. on Composite Materials* (Paris) 5–9 July 1999
[24] De Morais A B and Pereira A B 2007 Application of the effective crack method to mode I and mode II interlaminar fracture of carbon/epoxy unidirectional laminates *Composites A* **38** 785–94
[25] Bonhomme J, Arguelles A, Vina J and Vina I 2009 Fractography and failure mechanisms in static mode I and mode II delamination testing of unidirectional carbon reinforced composites *Polym. Test.* **28** 612–7
[26] Rased M F A and Yoon S H 2021 Experimental study on effects of asymmetrical stacking sequence on carbon fiber/epoxy filament wound specimens in DCB, ENF, and MMB tests *Compos. Struct.* **264** 113749

[27] Quan D, Deegan B, Alderliesten R, Dransfeld C, Murphy N, Ivankovic A and Benedictus R 2020 The influence of interlayer/epoxy adhesion on the Mode-I and Mode-II fracture response of carbon fibre/epoxy composites interleaved with thermoplastic veils *Mater. Des.* **192** 108781
[28] Al-Khudairi O, Hadavinia H, Waggott A, Lewis E and Little C 2015 Characterising mode I/ mode II fatigue delamination growth in unidirectional fibre reinforced polymer laminates *Mater. Des.* **66** 93–102
[29] Rikards R, Buchholz F G, Bledzki A K, Wacker G and Korjakin A 1996 Mode I, mode II, and mixed-mode I/II interlaminar fracture toughness of GFRP influenced by fiber surface treatment *Mech. Compos. Mater.* **32** 439–62
[30] Cintra G G, Vieira J D, Cardoso D C T and Keller T 2023 Mode I and mode II fracture behavior in pultruded glass fiber-polymer–experimental and numerical investigation *Composites B* **266** 110988
[31] Sampath P S, Murugesan V, Sarojdevi M and Thanigaiyarasu G 2008 Mode I and mode II delamination resistance and mechanical properties of woven glass/epoxy-PU IPN composites *Polym. Compos.* **29** 1227–34
[32] Joar M, Low K O and Wong K J 2018 Mode I and mode II delamination of a chopped strand mat E-glass reinforced vinyl ester composite *Plast. Rubber Compos.: Macromol. Eng.* **47** 391–7
[33] Dharmawan F, Simpson G, Herszberg I and John S 2006 Mixed mode fracture toughness of GFRP composites *Compos. Struct.* **75** 328–38
[34] Stavanovic D, Jar P Y B, Kalyanasundaram S and Lowe A 2000 On crack-initiation conditions for mode I and mode II delamination testing of composite materials *Compos. Sci. Technol.* **60** 1879–87
[35] Twelvetrees W N 1969 The evolution of the rotor blade: a review of helicopter blade construction with a survey of possible future developments in this field *Aircr. Eng. Aerosp. Technol.* **41** 19–23
[36] Brocklehurst A and Barakos G N 2013 A review of helicopter rotor blade tip shapes *Prog. Aerosp. Sci.* **56** 35–74
[37] Shi S Q 2007 Diffusion model based on Fick's second law for the moisture absorption process in wood fiber-based composites: is it suitable or not? *Wood Sci. Technol.* **41** 645–58
[38] Robinson R44 Maintenance Manual. Robinson Technical Publications, Robinson Helicopter Company Inc., Torrance, CA, USA 2021
[39] Amura M, Aiello L and Colavita M 2014 Failure of a helicopter main rotor blade *Proc. Mater. Sci.* **3** 726–31
[40] Air transportation safety investigation report A19Q0109: main rotor blade failure in flight, Robinson R44 (helicopter), C-FJLH, Lac Valtrie, Quebec. Transportation and Safety Board of Canada, 10th July 2019 https://www.bst-tsb.gc.ca/eng/rapports-reports/aviation/2019/a19q0109/a19q0109.html [date seen: 12.06.2024]
[41] Rasuo B 2009 Full-scale fatigue testing of the helicopter blades from composite laminated materials in the development process *J. Mech. Behav. Mater.* **19** 331–9
[42] Lombardi F 2017 *Building a Blade* (Rockville, MD: Access Intelligence)
[43] Pascal F, Navarro P, Marguet S and Ferrero J F 2017 Study of medium velocity impacts on the lower surface of helicopter blades *Dynamic Response and Failure of Composite Materials and Structures* (Cambridge: Woodhead Publishing) 159–81 pp (an imprint of Elsevier), The Officers' Mess Business Centre, Royston Road, Duxford, CB22 4QH, United Kingdom

[44] Amadasi L, Amadasi A, Buschmann C T and Tsokos M 2022 Fatal injuries due to direct helicopter propeller strike *Forensic Sci. Med. Pathol.* **18** 545–8

[45] Yu G, Li X and Huang W 2024 Performance and damage study of composite rotor blades under impact *Polymers* **16** 623

[46] Yeo H, Bousman W G and Johnson W 2002 Performance analysis of a utility helicopter with standard and advanced rotors *Presented at the American Helicopter Society Aerodynamics, Acoustics, and Test and Evaluation Technical Specialist Meeting* (San Francisco, CA, January 23–25 2002)

[47] Shinoda P M, Norman T R, Jacklin S A, Yeo H, Bernhard A P F and Haber A 2004 Investigation of a full scale wide chord blade rotor system in the NASA Ames 40- by 80-foot wind tunnel *Presented at the American Helicopter Society 4th Decennial Specialist's Conference on Aeromechanics* (San Francisco, CA) January 21–23 2004

[48] Thomas J 2022 Investigation completed for the two Utah Army National Guard UH-60 Black Hawks involved in training accident. Utah National Guard Official Department of Defense Website https://ut.ng.mil/NEWS/Article/3024354/investigation-completed-for-the-two-utah-army-national-guard-uh-60-black-hawks/ [date seen: 10.06.2024]

[49] Zimmerman N and Wang P H 2020 A review of failure modes and fracture analysis of aircraft composite materials *Eng. Fail. Anal.* **115** 104692

[50] Rajesh S, Suresh G and Mohan R C 2017 A review on material selection and fabrication of composite solid rocket motor (SRM) casing *Int. J. Mech. Solids* **12** 125–38

[51] Benli S and Sayman O 2011 The effects of temperature and thermal stresses on impact damage in laminated composites *Math. Comput. Appl.* **16** 392–403

[52] Hossam I, Saleh S and Kamel H 2019 Review of challenges of the design of rocket motor case structures *IOP Conf. Ser.: Mater. Sci. Eng.* 610 012019

[53] Lee H and Alam P 2021 The design of carbon fibre composite origami airbrakes for endeavour's Darwin I rocket *J. Compos. Sci.* **5** 147

[54] Xue J and Kirane K 2022 Cylindrical microplane model for compressive kink band failures and combined friction/inelasticity in fiber composites II *Compos. Struct.* **291** 115589

[55] Takahasi T, Ueda M, Iizuka K, Yoshimura A and Yokozeki T 2019 Simulation on kink-band formation during axial compression of a unidirectional carbon fiber-reinforced plastic constructed by X-ray computed tomography images *Adv. Compos. Mater.* **28** 347–63

[56] Liu Z, Hui W, Chen G and Cao P 2023 Multiscale analyses of the damage of composite rocket motor cases *Frontiers Mater.* **10** 1198493

[57] Zhang Z, He M, Liu A, Singh H K, Ramakrishnan K R, Hui D, Shankar K and Morozov E V 2018 Vibration-based assessment of delaminations in FRP composite plates *Composites B* **144** 254–66

[58] Srivastava L, Krishnanand L, Behera S and Nath N K 2022 Failure mode effect analysis for a better functional composite rocket motor casing *Mater. Today: Proc.* **62** 4445–54

[59] Kumar R and Kumar B 2024 Multiple reinforced composites, the possible solutions to avoid delamination in composite rocket motor casings *Springer Proc. Mater.* **39** 51–9

[60] Martin P J 1972 Failure analysis of solid rocket apogee motors *Final Report (NASA-CR-1228354) Standford Research Institute, California Institute of Technology, Pasadena, California 91 103, USA*

[61] NASA Independent Review Team (2018) Space X CRS-7 Accident Investigation Report, 12 March 2018, National Aeronautics and Space Administration, https://www.nasa.gov/wp-content/uploads/2018/03/public_summary_nasa_irt_spacex_crs-7_final.pdf?emrc=deb5c4 [date seen: 19.06.2024]

[62] Castavecchi D 2023 Europe's backlog of space missions worsened by rocket woes: Vega C launch failure, tracked to a Ukraine-made part, could further delay a handful of missions *Nature* (News Article): 3rd March 2023 https://www.nature.com/articles/d41586-023-00672-3 [date seen: 19.06.2024]

IOP Publishing

Composite Interfaces in Mechanical Design

Parvez Alam

Chapter 2

Chemical characterisation of fibre interfaces by FTIR

2.1 Introduction

Fourier transform infra-red (FTIR) spectroscopy is an analytical method involving the passing of IR radiation through a sample. While some of the IR radiation is absorbed by the molecules in the sample, some may be transmitted. A spectrum is formed representing molecular transmission and absorption of the IR radiation. The spectrum is often considered analogous to a fingerprint, as no two different materials will exhibit the same spectrum. As such, FTIR can be used to identify materials, to characterise materials in terms of their chemical constituents, and to determine the consistency or homogeneity of a material.

Particulate or fibre reinforcements added to a polymer matrix can have a number of effects at the molecular level that may vary the absorption/transmission FTIR 'spectral fingerprint'. Not only is there the obvious variation in chemical constituents, but the bonding between reinforcement and matrix will determine the strength and extent of molecular pinning [1], or immobility, at the interfaces, and the depth of the resultant interphase from the reinforcement surfaces, which itself will sometimes have renewed molecular orientations and mobilities, and may exhibit detectable chemical variants from the original. Several additional factors may have their own effects, such as reinforcement spacing, the original polymer crystallinity/amorphicity ratio, the molecular weight of the polymer, the sizes and size distributions of the reinforcements, and more. Each variable will alter the absorption and transmission spectrum, provided there are *detectable* chemical changes. Sizing types and changes to the surface chemistry as a result of treating or sizing reinforcements are detectable quite easily by FTIR, and as such, there are several works focussing on the use of FTIR spectroscopy to determine the chemistry of fibre surfaces.

doi:10.1088/978-0-7503-5688-6ch2

This chapter aims to introduce examples of FTIR wavenumbers and bands relevant to some common engineering reinforcing fibres. These include carbon, glass, aramid, and natural fibres classes. The chapter will cover example expected wavenumbers and bands relevant to the untreated fibres, as well as expected variations arising from different surface treatments applied to the different classes of fibre discussed.

2.2 Carbon fibres

2.2.1 PAN-based carbon fibres

With certain fibres such as polyacrylonitrile (PAN)-based carbon fibres (CFs), FTIR can successfully identify whether fibre stabilisation is complete or not. PAN is an organic resin, semicrystalline in nature, which is a precursor to thermally stabilised, high-quality CFs. While PAN fibres are themselves thermally oxidised/stabilised at elevated temperatures (e.g. 200 °C–300 °C), to manufacture a high-quality CF, the stabilised PAN fibre should be carbonised at temperatures at or exceeding 1000 °C. The stabilisation process already removes many of the chemical characteristics of PAN fibres as shown in figure 2.1 (from [2]). Unlike CFs, PAN fibres exhibit carbon–nitrogen triple bonds, [C≡N], carbon–oxygen double bonds, [C=O], carbon–oxygen–carbon bonds, [C—O—C], carbon–hydroxide bonds, [C—OH], doubled-bonded carbon to single-bond hydrogen bonds, [=C—H], and carbon–hydrogen bonds, [C—H], each of which will show distinctive, sharp peaks relative to carbonised PAN fibres or stabilised PAN. Carbonised fibres as well as stabilised PAN may show [O—H], [C—H] peaks, [C≡N] peaks, and [C=N], [C=C], and [N—H] peaks (mixed), which are relative peaks and may vary based on the temperature and time of carbonisation. Some of these peaks also appear in sizing to coat CFs as sizing is typically used to improve interfacial bonding and to protect the fibres during the handling process [3].

2.2.2 Surface treated and sized CF

Sizing chemistry can sometimes be seen as a 'black box' outside of industry, and while there are chemical differences in the sizing used for CF, recognisable relative differences from carbonised CF will be important when identifying the 'spectral fingerprint' of the sizing used for a particular carbon fibre. Comparing sized Hexcel IM7 and Toray T800HB fibres, for example, Brocks and co-workers [4] identified commonalities in terms of the presence of [N—H] stretching (3300–2300 cm^{-1}), ester groups (2278 cm^{-1}), carboxylates (1705 cm^{-1}), and both common [C—H] stretch (1045 cm^{-1}) and deformational (1425 cm^{-1}, 824 cm^{-1}, 801 cm^{-1}) peaks. Notable differences were reported as appearing from aromatic nitrile compounds (1501 cm^{-1}) and from [C—S] stretches (1178 cm^{-1}), each of which was only present on the T800HB fibres. Notably, [C—C] stretches at (1234 cm^{-1}) were present in both IM7 and T800HB. The type of sizing can also be identified using FTIR with epoxy type sizing revealing peaks at or close to (915 cm^{-1}) as reported by Dai and co-workers [5] who compared specifically Toray T300B and T700SC CFs.

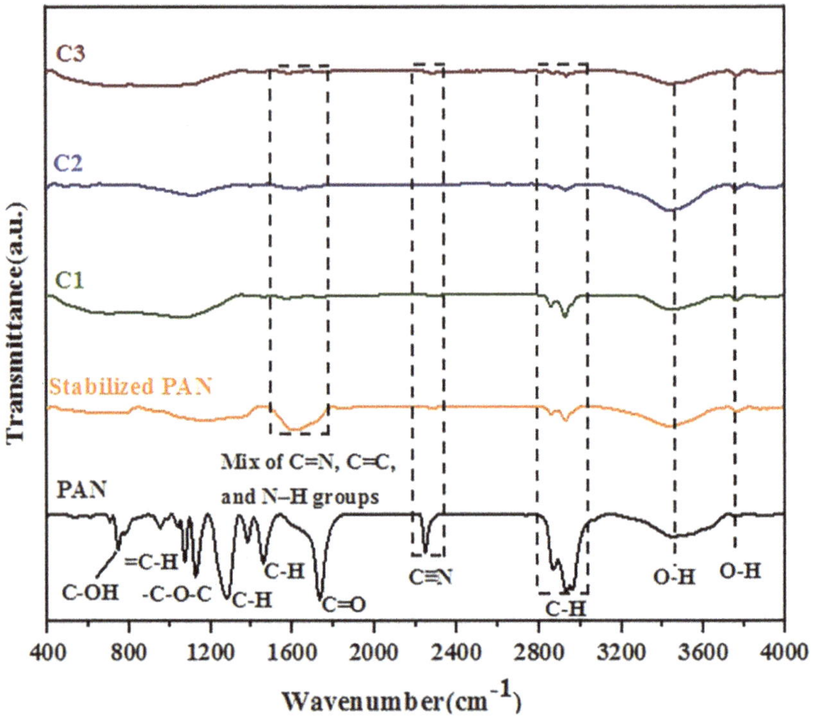

Figure 2.1. FTIR spectra of fibre samples in the range 400–4000 cm^{-1}. Reproduced from [2], CC BY 4.0. Here PAN refers to unstabilised PAN fibres. Stabilized PAN has been heated at 300 °C with a heating rate of 5 °C min^{-1} for 1 h in air, and C1, C2,and C3 curves represent stabilised PAN fibres carbonised at 1000 °C, 1200 °C, and 1400 °C, respectively, for 1 h under a nitrogen gas flow at a heating rate of 10 °C min^{-1}.

In addition to the differentiation of types of sizing, FTIR can be used to differentiate the resultant surface chemistry of fibres subjected to specific surface treatments. Ma and co-workers [6], for example, used a variety of chemical surface modifications to base carbon fibre and used FTIR to reveal the individual spectra of the surface modified fibres, figure 2.2. As seen in the figure, the addition of carboxyl groups of CF surfaces to form CF-COOH yields new [C═O] stretching vibrations at 1741 cm^{-1} and an [—OH] stretching vibration at 3380 cm^{-1}. MXene ($Ti_3C_2T_x$) is understood to interact with the coated CF-COOH in CF-COOH/MXene and as such the peaks are similar to the CF-COOH peaks, while the CF-CONH-MXene spectrum shows [—N—H] bending vibration at 1587 cm^{-1} and a [—C═O] stretching vibration at 1721 cm^{-1}, which is indicative of the surface presence of [CONH—] and the covalent bonding of MXene to the CF surface.

2.2.3 Summary table

Example FTIR peaks and bands are summarised in table 2.1 for treated and untreated CFs.

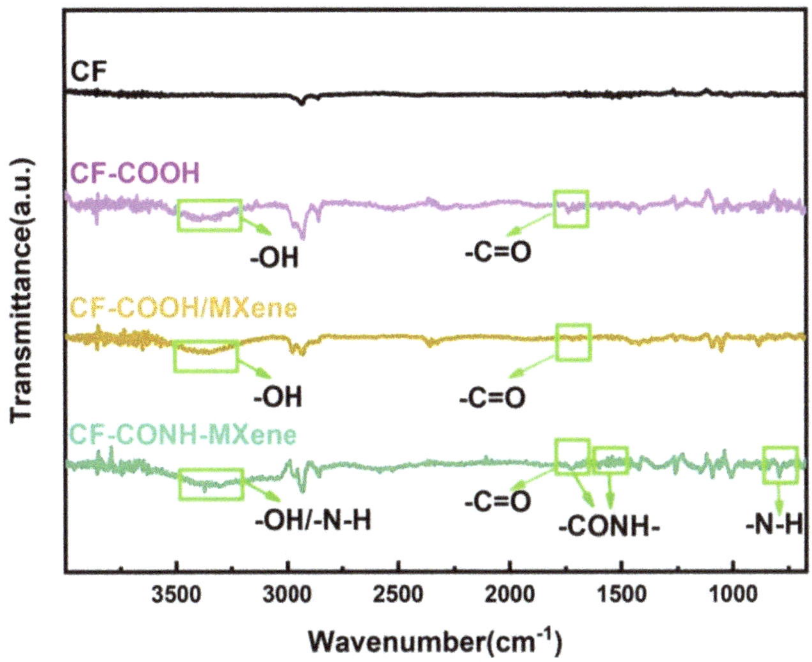

Figure 2.2. FTIR spectra of CF, CF-COOH, CF-COOH/MXene, and CF-CONH-MXene. Reprinted from [6], Copyright (2022), with permission from Elsevier. Here CF = carbon fibre, CF-COOH = carbon fibre surface modified with carboxyl groups, CF-COOH/MXene = carbon fibre surface modified with both carboxyl groups and MXene ($Ti_3C_2T_x$), and CF-CONH-MXene = carbon fibre with series chains of CONH-MXene attached to the fibre surface.

2.3 Glass fibres

2.3.1 Treated and untreated glass fibres

Glass or silica fibre chemistry differs based on the type of glass, the type of fibre treatment applied, or the original source of the silica [7]. Untreated E-glass fibre, for example, will typically display a peak near 900 cm^{-1} indicating an [Si—O—Si] bond [8], a peak that is also noticeable in basalt fibres, which are made from igneous rocks and thus contain similar [Si—O—Si] peaks.

Borosilicate glasses contain boron trioxide, an addition that enables very low thermal expansion in the glass fibre. Peaks for fibres made of these glasses are reported at 767 cm^{-1} for [Si—O—Si] and 918 cm^{-1} for [Si—O] bonds [9], which when treated with polydimethylsiloxane to generate surface hydrophobicity will show an additional peak at 1008 cm^{-1} indicating new [Si—O—Si] bonds from the presence of the siloxane treatment, and a peak at 1259 cm^{-1} indicating the presence of a new [Si—CH_3] functional group.

Silanes are often used as glass fibre sizing and silanised glass fibres are expected to show peaks at 828 cm^{-1} representing the stretching of [Si—OCH_3] due to the presence of silane [10]. Amino silanes such as 3-aminopropyltrimethoxysilane

Table 2.1. A few example FTIR peaks in treated and untreated carbon fibres.

Wavenumber (cm^{-1})	Association
3380	—OH stretching through the addition of carboxyl groups —COOH to carbon fibre surfaces [6]
3300–2300	N—H stretching common to both Hexcel IM7 and Toray T800HB fibres (including sizing) [4]
2278	Ester groups common to both Hexcel IM7 and Toray T800HB fibres (including sizing) [4]
1741	C═O stretching through the addition of carboxyl groups —COOH to carbon fibre surfaces [6]
1721	C═O stretching due to the presence of —CONH and the covalent bonding of MXene ($Ti_3C_2T_x$) on carbon fibre surfaces [6]
1705	Carboxylates common to both Hexcel IM7 and Toray T800HB fibres (including sizing) [4]
1587	—N—H bending through the addition of —CONH—MXene to carbon fibre surfaces [6]
1501	C—H deformational common to both Hexcel IM7 and Toray T800HB fibres (including sizing) [4]
1425	Aromatic nitrile compounds in Toray T800HB fibres (including sizing) [4]
1234	C—C stretching common to both Hexcel IM7 and Toray T800HB fibres (including sizing) [4]
1178	C—S stretching in Toray T800HB fibres (including sizing) [4]
1045	C—H stretching common to both Hexcel IM7 and Toray T800HB fibres (including sizing) [4]
915	Epoxy type sizing common to both Toray T300B and T700SC fibres [5]
824	C—H deformational common to both Hexcel IM7 and Toray T800HB fibres (including sizing) [4]
801	C—H deformational common to both Hexcel IM7 and Toray T800HB fibres (including sizing) [4]

(APTMS) [10] and (3-aminopropyl) triethoxysilane (APTES) [8] are often used as silane coupling agents, and as such, when fibres are silanised, they will show additional peaks related to, for example, [—CH] stretching of the propyl groups present at 2900 cm^{-1}, 2915 cm^{-1} and 3003 cm^{-1}, [—CH] bending of the propyl groups present at 1490 cm^{-1}, and [NH_2] stretching due to the presence of amine groups in amino silanes at 3432 cm^{-1} [8, 10].

Further processes to siliconise fibre surfaces after the addition of silane groups should give rise to a further peak at 998 cm^{-1} representing the stretching of [Si—O—Si] post-siliconisation. The general range within which [Si—O—Si] can be broad and the specific peak will depend on the treatment, with many of the peaks appearing between 900 and 1200 cm^{-1}.

2.3.2 Summary table

Example FTIR peaks and bands are summarised in table 2.2 for treated and untreated glass fibre.

2.4 Aramid fibres

Aramids are aromatic polyamides. These are strong, tough, and heat-resistant advanced materials used in high-end applications such as aerospace and as military textiles. There are a number of trademarked aramid fibres, which can be subdivided into those that are m-aramids (poly(m-phenylene isophthalamide)) including Nomex® (DuPont Co.) and Conex® (Teijin Ltd), and those that are p-aramids (poly(p-phenylene terephthalamide)) including Kevlar® (DuPont Co.) and Twaron® (Akzo Industrial Fibers) [3].

2.4.1 M-aramids and p-aramids

Observable FTIR peaks between m-aramids and p-aramids tend to be relatively similar as noted by [11], with absorption peaks at 3314 cm^{-1} attributable to the stretching of [N—H] bonds, and stretching vibrations at 2920 cm^{-1} and 2850 cm^{-1} from the presence of [$C—H_2$] and [$C—H_3$] groups. [N—H] deformation and [C—N] stretching are observed at 1532 cm^{-1} and 2850 cm^{-1}, respectively, and peaks observed at 1507 cm^{-1} and 1307 cm^{-1} are from the stretching vibration of the [C=C] skeleton on a benzene ring and the amide III band, respectively. The amide III band

Table 2.2. A few example FTIR peaks in treated and untreated glass fibres.

Wavenumber (cm^{-1})	Association
3432	NH_2 stretching of amine groups in amino silanes due to the silanising of glass fibres [8, 10]
3003	—CH stretching of propyl groups in silanised glass fibres [8, 10]
2915	—CH stretching of propyl groups in silanised glass fibres [8, 10]
2900	—CH stretching of propyl groups in silanised glass fibres [8, 10]
1490	—CH bending of propyl groups in silanised glass fibres [8, 10]
1259	$Si—CH_3$ functional group following the polydimethylsiloxane (PDMS) treatment of borosilicate glass fibres [9]
1200–900	General broad range possible within which Si—O—Si stretching peaks may appear [10]
1008	Si—O—Si bonding from the presence of siloxane after polydimethylsiloxane (PDMS) treatment of borosilicate glass fibres [9]
998	Si—O—Si stretching of siliconised glass fibres [10]
900	Si—O—Si bonding in untreated E-glass and balast fibres [8]
918	Si—O bonding in borosilicate glass fibres [9]
828	$Si—OCH_3$ stretching in silanised glass fibres due to the presence of silane [10]
767	Si—O—Si bonding in borosilicate glass fibres [9]

consists of [C—N] stretching, [N—H] bending, and [C—C] stretching vibrations. Xu and co-workers [11] note that m-aramids have greater IR absorption (indicated by a larger area under the curve) at the 4000–3300 cm^{-1} and 1600–1200 cm^{-1} bands than p-aramids, and that this may be due to the groups being weakly restricted (due to being less crystalline) in the m-aramids than the p-aramids.

2.4.2 Types of aramid fibre

IR differences can also be observed between different types of aramid fibres belonging to the same group. Shebanov and co-workers [12], for example, discuss differences between p-aramid fibres Technora® (Teijin Ltd) T200w and T240 and Twaron® 1000 (Teijin Ltd) drawing from the works of Derombise and co-workers [13, 14]. In their paper, they note that amide III bands occur at 1303 cm^{-1} and 1306 cm^{-1} in T200w and T240 fibres, respectively, and at 1305 cm^{-1} in the Twaron 1000 fibres. Amide II bands were noted to occur at 1639 cm^{-1} in both T200w and T240, but at 1538 cm^{-1} in Twaron 1000. While these peaks seem to differentiate the Technora fibres from the Twaron fibre, the full IR fingerprint reveals the [C—H] aromatic rings, which peak at 820 cm^{-1} in T200w and Twaron 1000, dissimilarly to the peak in T240, which is noticeable at 824 cm^{-1}.

2.4.3 Surface-treated aramid fibres

Surface treatments have also been shown to be detectable on aramids through FTIR spectroscopy. For example, Kevlar® (DuPont Co.) can be surface functionalised using NaOH and HCl treatments in series [15]. Here, the chemical structure of Kevlar® (DuPont Co.) without treatment is reported to show noticeable [—NH] stretching (amide link) at 3312 cm^{-1}, [—C═O] stretching at 1638 cm^{-1}, and [—NH] bending vibrations at 1534 cm^{-1}. The effect of NaOH and HCl surface treatment breaks the amide links, which can be recognised by the broadening of the 3300–3200 cm^{-1} amide peak due to the stretching of the primary amine [NH_2]. Thirty percent phosphoric acid treatment at 40 °C to Twaron2200® (Teijin Ltd) fibres shows the characteristic aramid absorption peaks for [—NH] stretching (amide link) at 3320 cm^{-1} and [—NH] bending vibrations at 1543 cm^{-1} [16]. When the aramid fibres are further functionalised using a SiO_2/polyurethane, the specific components of this hybrid coating can be recognised through polyurethane peaks at 2958 cm^{-1}, 1610 cm^{-1}, and 1222 cm^{-1}, while an adsorption peak at 1067 cm^{-1} is associated with the presence of the [Si—O—Si] [16]. Aramid sizing can also be applied through grafting techniques, several of which are conducted in supercritical carbon dioxide (scCO_2). scCO_2 is essentially a fluidic state of CO_2, where it is held above its critical temperature (30.978 °C) and pressure (7.3773 MPa). The process has the benefit of enabling fibre bulging and fibre surface to subsurface restructuring such that treatments can more effectively penetrate the fibre, as opposed to merely sitting atop the fibre surface. Importantly though, damage to fibres subjected to scCO_2 can be detected through FTIR, an example of which is discussed by Li and co-workers [17], who report damage as being recognisable from the broadening of the hydrogen bond peak at 3310 cm^{-1}. The grafting of glycidyl-polyhedral oliomeric silsesquioxane in

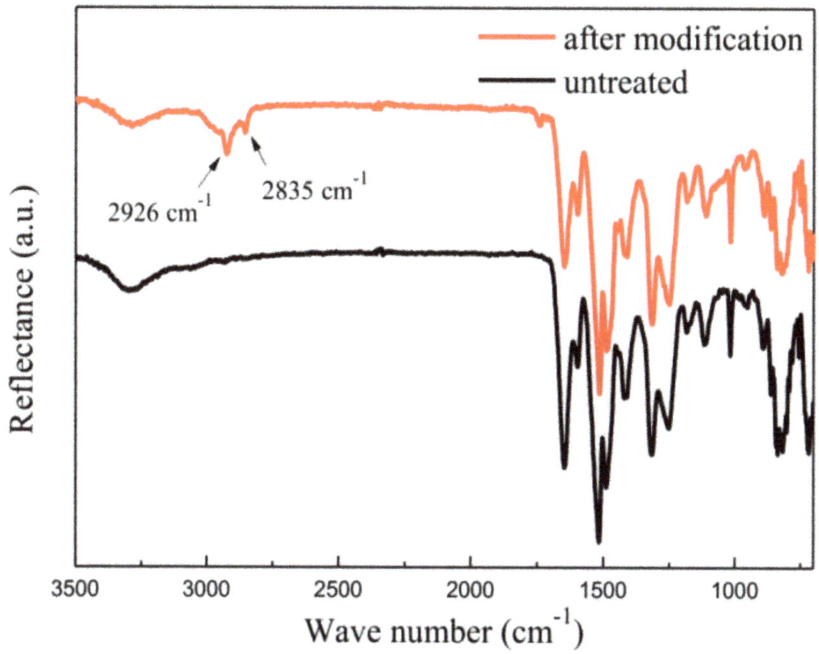

Figure 2.3. FTIR spectra of untreated aramid fibre (black line) and aramid fibre grafted with 1,4-dichlorobutane in $scCO_2$ (red line). Reproduced from [18]. CC BY 4.0.

$scCO_2$ can in turn be recognised by the development of peaks at 2850 cm^{-1} and 2920 cm^{-1}. These appear due to the stretching of both [–CH–] and [–CH_2–] groups, which do not appear in untreated aramid fibres (AF-1000, 1500D by South Alkex Company, Seoul, Korea). Grafting 1,4-dichlorobutane to aramid fibres (sourced from China Aerospace Science and Technology Group 46th Research Institute (Huhhot, China)) in $scCO_2$ shows similar [–CH–] and [–CH_2–] stretching vibrations at 2835 cm^{-1} and 2926 cm^{-1}, respectively [18], figure 2.3.

2.4.4 Summary table

Example FTIR peaks and bands are summarised in table 2.3 for treated and untreated aramid fibre.

2.5 Natural fibres

Natural fibres used in biocomposites are predominantly plant based and thus comprise cellulose, hemicelluloses, and lignin. Natural fibres applied to composites may include bast fibres such as flax, hemp, jute, kenaf, and ramie; seed fibres including cotton, kapok, and coconut; leaf fibres including sisal and pineapple; and grass/reed fibres including corn, rice, and wheat [19].

Table 2.3. A few example FTIR peaks in treated and untreated aramid fibres.

Wavenumber (cm^{-1})	Association
4000–3300	Band broader in m-aramid than in p-aramid [11]
3320	Twaron2200 fibre manufactured by Teijin Ltd treated in 30% phosphoric acid at 40 °C [NH] stretch [16]
3310	Broadening of hydrogen bond peak caused by aramid fibre damage and restructuring when subjected to $scCO_2$ (using AF-1000, 1500D manufactured by South Alkex Company, Seoul, Korea) [17]
3300–3200	Band broader in Kevlar fibre manufactured by DuPont Co. subjected to NaOH and HCl treatment due to the primary amine [NH_2] stretch [16]
3316	[CH] stretch derived from hydrogen bond association state in aramid fibre (AF-1000, 1500D) manufactured by South Alkex Company, Seoul, Korea [17]
3314	Aramid fibre [NH] stretch [11]
3312	Kevlar fibre manufactured by DuPont Co. [NH] stretch [16]
2958	Aramid fibre SiO_2/polyurethane sized peak associated with polyurethane component of sizing [16]
2926	Aramid fibre (sourced from China Aerospace Science and Technology Group 46th Research Institute (Huhhot, China)) grafted with 1,4-dichlorobutane in $scCO_2$, [—CH_2—] stretch [18]
2920	Aramid fibre [CH_2] stretch [11]
2920	[CH] vibration from glycidyl-polyhedral oliomeric silsesquioxane in $scCO_2$ grafted to aramid fibre (AF-1000, 1500D) manufactured by South Alkex Company, Seoul, Korea [17]
2850	[CH_2] vibration from glycidyl-polyhedral oliomeric silsesquioxane in $scCO_2$ grafted to aramid fibre (AF-1000, 1500D) manufactured by South Alkex Company, Seoul, Korea [17]
2850	Aramid fibre [CH_3] stretch [11]
2850	Aramid fibre [CN] stretch [11]
2835	Aramid fibre (sourced from China Aerospace Science and Technology Group 46th Research Institute (Huhhot, China)) grafted with 1,4-dichlorobutane in $scCO_2$, [—CH—] stretch [18]
1639	Technora®: (Teijin Ltd) T200w and T240 fibre Amide II [12–14]
1638	Kevlar fibre manufactured by DuPont Co. [C═O] stretch [16]
1636	Stretching vibration of [C═O] Amide I band in aramid fibre (AF-1000, 1500D) manufactured by South Alkex Company, Seoul, Korea [17]
1610	Aramid fibre SiO_2/polyurethane sized peak associated with polyurethane component of sizing [16]
1600–1200	Band broader in m-aramid than in p-aramid [11]
1543	Twaron2200 fibre manufactured by Teijin Ltd treated in 30% phosphoric acid at 40 °C [NH] bending [16]
1540	curved vibration of [—N—H] in aramid fibre (AF-1000, 1500D) manufactured by South Alkex Company, Seoul, Korea [17]

(*Continued*)

Table 2.3. (*Continued*)

Wavenumber (cm^{-1})	Association
1538	Twaron®: (Teijin Ltd) 1000 fibre Amide II [12–14]
1534	Kevlar fibre manufactured by DuPont Co. [NH] bending [16]
1532	Aramid fibre [NH] stretch [11]
1507	Aramid fibre [C=C] stretch (benxene ring) [11]
1307	bending vibration of [–N–H] in aramid fibre (AF-1000, 1500D) manufactured by South Alkex Company, Seoul, Korea [17]
1307	Aramid fibre [C=C] stretch Amide III band [11]
1306	Technora®: (Teijin Ltd) T240 fibre Amide III [12–14]
1305	Twaron®: (Teijin Ltd) 1000 fibre Amide III [12–14]
1303	Technora®: (Teijin Ltd) T200w fibre Amide III [12–14]
1222	Aramid fibre SiO_2/polyurethane sized peak associated with polyurethane component of sizing [16]
1067	Aramid fibre SiO_2/polyurethane sized peak associated with stretching of [Si–O–Si] component of sizing [16]
824	Technora®: (Teijin Ltd) T240 fibre [C–H] aromatic rings [12–14]
820	Technora®: (Teijin Ltd) T200w and Twaron®: (Teijin Ltd) 1000 fibre [C–H] aromatic rings [12–14]

2.5.1 Characterisation of natural fibre constituents

The structural backbone of plant fibres is cellulose, a predominantly crystalline polysaccharide that provides stiffness and strength to the fibre. A summary of the characteristic wavenumbers (and bands) for lignocellulosic materials compiled by [20] from [21, 22, 24] informs that [O–H] stretching in free or weakly hydrogen bonded hydroxyls will appear between 3550 and 3650 cm^{-1}, while [O–H] from hydroxyls that are not weakly bonded appear between 3200 and 3400 cm^{-1}. Wavenumbers appearing between 2840 and 2940 cm^{-1} are characteristic of [C–H] stretching while [C=O] stretching in carbonyls are expected to be seen between 1720–1740 cm^{-1}. [C–H] deformations due to the methoxylate group in lignins are expected to be seen between 1400 and 1430 cm^{-1}, while other lignin-based peaks include [C–O] stretching in lignin at 1327 cm^{-1} (syringyl ring) and between 1250 and 1260cm^{-1} (guaiacyl ring), [C=C] symmetrical stretching of aromatics at 1506 cm^{-1} [23], [O–H] deformation in lignin due to the syringyl structure at 1230 cm^{-1}, [C=O] and G ring stretching at 1246 cm^{-1} [23], and [C–H] out-of-plane vibrations owing to the lignin at 830 cm^{-1}. A peak similar to syringyl ring structure at 1327 cm^{-1} can also arise in relation to [CH_2] wagging in cellulose. Peaks between 1150 and 1160 cm^{-1} can be associated with [C–O–C] antisymmetrical stretching in cellulose and aromatic [C–HCH_2] wagging in cellulose. The [C–O] stretching of p glycosidic linkages in cellulose can be observed between 1098 and 1120 cm^{-1}, while [C–O] stretching in cellulose has characteristic peak at 1036 cm^{-1}, though it should be noted that this peak also corresponds to aromatic

[C—H] deformations (guaiacyl structure) and [C—O] deformations of the primary alcohols in lignin. [C—O] stretching in cellulose is also expected at 1003 cm^{-1} and antisymmetric stretching due to the presence of β linkages in cellulose are between 890 and 900 cm^{-1}. [C—O] bond stretching can also be noted owing to the acetyl groups in xylan and hemicelluloses between 1240 and 1245 cm^{-1} [20].

2.5.2 Characterisation of cellulose polymorphs

Two polymorphs of cellulose (I and II) can be differentiated using FTIR spectroscopy, and it should be noted that these polymorphs are the most studied out of the four possible polymorphs of cellulose. Cellulose I is comprised of parallel cellulose strands but without intersheet bonding, while cellulose II is comprised of antiparallel strands with intersheet bonding and is consequently thermodynamically more stable than cellulose I. In cellulose I, there are two intramolecular hydrogen bonds (at O(2)H–O(6) and at O(3)H–O(5)) and one intermolecular bond at O(6)H–O(3). Cellulose II contains three intramolecular bonds (O(2)H–O(6), O(3)H–O(5), and O (2)H–O(2)) and two intermolecular bonds (O(6)H–O(2) and O(6)H–O(3)) [25]. Cellulose I is naturally occurring while cellulose II can form after surface treatment. One example is through mercerisation treatment with NaOH, which is a method used to strengthen natural fibres, to improve surface hydrophobicity, and to reduce shrinkage, each of which can be a desired fibre property for use in structural composites and in textiles manufacture. These polymorphs I and II of cellulose are shown in figure 2.4, with cellulose I shown in (a) and cellulose II shown in (b) (from [26]). In this image, the 020 plane is represented, and green-coloured atoms are assumed to be water binding, with water-binding peaks being evident at 1625–1660 cm^{-1} (adsorbed water on non-crystalline cellulose) and at ca. 1600 cm^{-1} (vibrations due to adsorbed water). In the figure, the cellulose II, O(2)H–O(2′) hydrogen bond is

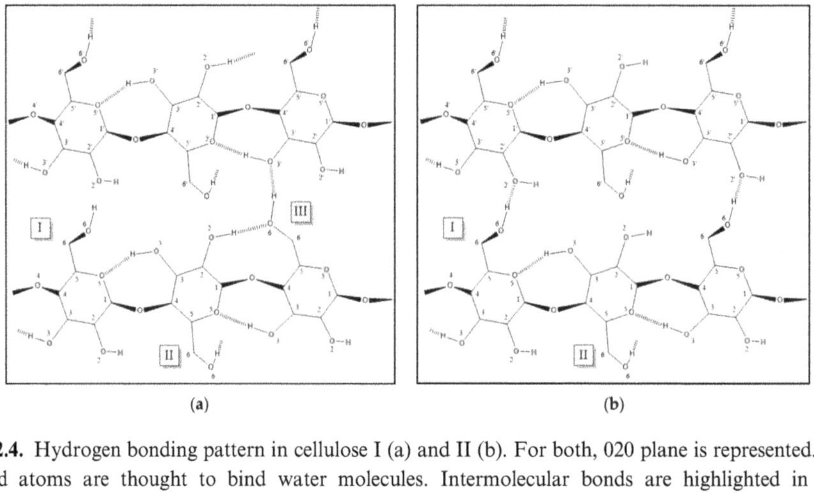

Figure 2.4. Hydrogen bonding pattern in cellulose I (a) and II (b). For both, 020 plane is represented. Green-coloured atoms are thought to bind water molecules. Intermolecular bonds are highlighted in red. In cellulose II, O(2)H–O(2′) hydrogen bond is not shown as it is visible on another plane (110). Roman numbers in labels stand for different conformations the C(6)O(6) group can assume in the same chain. Reproduced from [26], CC BY 4.0.

not shown as it is visible in the 110 plane, and Roman numbers in the labels indicate the different conformations the C(6)O(6) group is able to assume in the chain [26]. Fan and co-worker's [25] discussions on the work of [27, 28] elucidate a break down of critical IR peaks that enable a differentiation between the two polymorphs. Specifically, in cellulose I, O(6)H–O(3) hydrogen bonding has a band between 3230 and 3310 cm^{-1}, an O(3)H–O(5) band between 3340 and 3375 cm^{-1}, and an O(2)H–O(6) band between 3405 and 3460 cm^{-1}. In cellulose II critical peaks can be identified at 3175 cm^{-1} associated with [O—H] stretching, 3308 cm^{-1} associated with OH hydrogen bonds (intermolecular), 3309 cm^{-1} also associated with OH hydrogen bonds (intermolecular), and 3315 cm^{-1}, 3374 cm^{-1}, and 3486 cm^{-1} associated with OH hydrogen bonds (intramolecular).

2.5.3 Mercerised natural fibre

Mercerisation is an alkaline treatment that removes the lignin and hemicellulose components of natural fibres. What remains is predominantly cellulose, though there may also be residual waxes present [29]. Kapok fibres (*Aerva tomentosa*) mercerised in 5% NaOH for 7–8 h at room temperature are absent of lignin, lacking the peaks at 1240 cm^{-1} [30] characteristic of the [C—O] aryl group in lignin [31] and at 1740 cm^{-1}, which indicates a loss in the hemicellulose component of the fibre. This is also noted in [33] for mercerised coir fibres where the hemicellulose [C═O] peak expected at 1723 cm^{-1} is significantly decreased in intensity in the mercerised coir. In addition, the untreated kapok fibre exhibits a peak at 1630 cm^{-1} associated with the [—C═O] functional group [32], which after mercerisation, peaks at 1624 cm^{-1} [30]. Immersing coir fibres in 5% NaOH in water for 24 h reduces the hydrogen bonding from OH groups, which is essentially how the process of mercerisation improves the hydrophobic properties of fibre surfaces. This can also be recognised from the significant reduction in the intensity of the [—OH] band between 3000 and 2800 cm^{-1} as reported in [33]. The bands between 3100 and 3400 cm^{-1} associated with [O—H] stretching of hydroxyl groups present in hemicellulose will also be reduced in intensity in mercerised coir fibre [33]. [C═C] stretching of the aromatic ring in lignin peaks at 1591 cm^{-1} in untreated coir, and this is another peak that will be less obvious in the spectra for correctly mercerised coir fibre. A further reduced peak of mercerised coir fibre relative to untreated coir fibre is that associated with [C—O] stretching of the acetyl group of lignin ordinarily observed at 1293 cm^{-1}. *Typha domingensis* (southern cattail) treated with NaOH solution (6% w/v) for 7 h at room temperature is absent of similar peaks and bands associated with the functional groups for lignin and hemicellulose, but which appear in untreated southern cattail fibres at slightly different wavenumbers [34]. Here, reduced or missing absorbance peaks characteristic of surface hydroxyl groups are between 3332 and 2919 cm^{-1}, and [C═O] stretching of the amide groups in hemicellulose and lignin are found at 1507 cm^{-1} and 1361 cm^{-1}, respectively.

Fibres can also be cleaned of hemicellulose and lignin effectively using combined treatments. For example, fibres may be first dewaxed and delignified before mercerisation. Liew and co-workers [35], for example, used toluene and ethanol (2:1 v/v ratio) as dewaxing solvents, and acetic acid and hydrogen peroxide (in the presence of a titanium oxide catalyst) as delignifying solvents, prior to 6% NaOH mercerisation to remove hemicelluloses and pectins in both jute (*Corchorus olitorius*) and bamboo (*Dendrocalamus asper*) fibres. They compared the FTIR spectra of cellulose fibres subjected to these combined treatments against the original untreated fibres and against commercial cellulose as shown in figure 2.5, where (A) is the spectrum for untreated jute fibre, (B) is the spectrum for untreated bamboo fibre, (C) is the spectrum for commercial cellulose, (D) is the spectrum for treated extracted jute cellulose, and (E) is the spectrum for treated extracted bamboo cellulose. In this figure, the peak at 1738 cm^{-1} observed in untreated jute and bamboo fibres is associated with the acetyl and uronic ester groups or the ester linkage of carboxylic group of the ferulic and p-coumeric acids of hemicelluloses. This peak is absent in the extracted celluloses (D–E) and is indicative of the removal of hemicellulose. Additionally, the peak at 1242.4 cm^{-1} is associated with the stretching vibrations of phenolic hydroxyl groups in lignin, and while this peak is

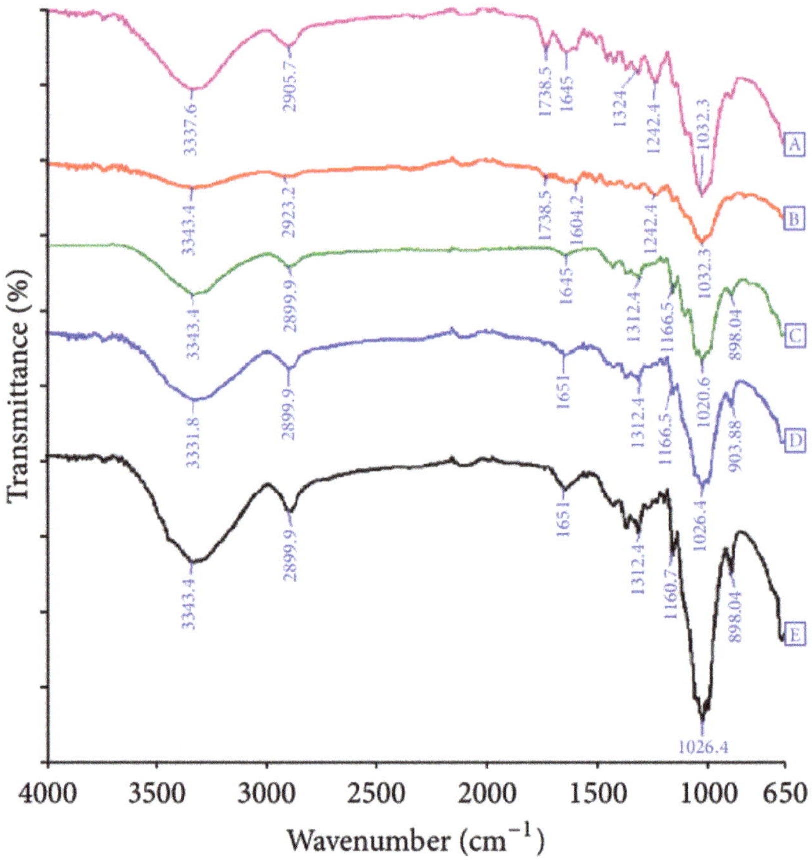

Figure 2.5. Wavenumbers of (A) untreated jute fibre, (B) untreated bamboo fibre, (C) commercial cellulose, (D) treated extracted jute cellulose, and (E) treated extracted bamboo cellulose [35]. John Wiley & Sons. CC BY 4.0.

observed in untreated jute and bamboo fibres, it is absent in treated extracted celluloses from these same fibres, indicating also the removal of lignin.

2.5.4 Acetylated natural fibre

Acetylation is a treatment that reduces the susceptibility of natural fibres to the absorption of moisture. The treatment usually involves either acetic acid and/or propionic acids, or their derived anhydrides, at elevated temperatures. Acetylation also improves the dimensional stability of fibres since swelling is reduced, and it further decreases the rate at which natural fibres degrade, a phenomenon otherwise accelerated by the presence of water as water improves the conditions for natural microbial activity. The treatment process is essentially an organic esterfication reaction of acetyl/propionyl groups and hydroxyl groups from the fibre surface (i.e. a nucleophilic acyl substitution). Tserki and co-workers [36] note that treatment of flax and hemp fibres (from S.A. Van Robaeys Freres, France), with acetic anhydride and propionyl anhydride, result in strong relative intensity IR peaks when compared against untreated fibres, at 1735 cm^{-1} (hemp) and 1737 cm^{-1} (flax). These high relative intensity peaks are associated with the esterfication of the hydroxyl groups and a stretching vibration of the carbonyl [C=O] group from the ester. In the untreated fibres, [C=O] stretching vibrations are also present except at 1740 cm^{-1}, which since coming from untreated fibres, are assumed to originate from the xylan component of the hemicellulose. It can be important to compare against the original untreated fibre as the IR peaks can be specific to the plant. Celino and co-workers [37], for example, associate absorption bands at 1735 cm^{-1} with [C=O] stretching of acetyl or carboxylic acids in flax and hemp fibres (as well as in sisal and jute), while the 1735 cm^{-1} band is considered associated with [C=O] vibrations from natural fibre pectins, which can be strengthened by carbonyl groups of oxycelluloses if the fibre is partially degraded [38]. Tserki and co-workers [36] also confirm esterification from the appearance of two new peaks at 1162/1163 cm^{-1} and 1229 cm^{-1}, which are absent in untreated fibres. These peaks are associated with [C–O] stretching from the ester carboxyl group and from acetates, respectively.

Carbonyl [C=O] stretching after acetylation in coir fibres is reported as occurring at the higher wavenumber of 1744 cm^{-1} [39] than previously discussed for hemp and flax fibres in [36]. Khalil and co-workers [39] additionally report that acetylated coir fibre will produce an increased peak at 1244 cm^{-1}, which is most likely to be associated with [C–O] stretching of acetyl groups from acetylation in addition to those expected of hemicellulose and pectin [37]. Similar band shifts have been reported for acetylated jute fibres from 1740 cm^{-1} to 1751 cm^{-1} ([C=O] stretch) and 1235 cm^{-1} to 1241 cm^{-1} ([C–O] stretch) [40]. While in both these cases of coir and jute fibre, acetylation treatment is reported to increase the wavenumber, the [C=O] stretching band at 1737 cm^{-1} decreases in the case of acetylated hemp fibres to 1726 cm^{-1} [41].

2.5.5 Summary table

Example FTIR peaks and bands are summarised in table 2.4 for treated and untreated natural fibre.

Table 2.4. A few example FTIR peaks in treated and untreated natural fibres.

Wavenumber (cm^{-1})	Association
3550–3650	[O—H] stretching in free or weakly hydrogen bonded hydroxyls [20–22, 24]
3450	Acetylated sisal fibres, [O—H] stretching [42]
3438	Acetylated sunhemp fibres, [O—H] stretching [42]
3434	NaOH treated hemp fibres, [O—H] stretching in cellulose [41]
3419	Acetylated hemp fibres, [O—H] stretching in cellulose [41]
3407	Untreated hemp fibres, [O—H] stretching in cellulose [41]
3405–3460	Cellulose I O(2)H–O(6) hydrogen bonding [25, 27, 28]
3340–3375	Cellulose I O(3)H–O(5) hydrogen bonding [25, 27, 28]
3386	Cellulose II intramolecular OH hydrogen bonds [25, 27, 28]
3374	Cellulose II intramolecular OH hydrogen bonds [25, 27, 28]
3370	Untreated sisal fibres, [O—H] stretching [42]
3332–2919	Untreated southern cattail fibres surface hydroxyl groups, absent in mercerised fibres [34]
3309	Cellulose II intermolecular OH hydrogen bonds [25, 27, 28]
3308	Cellulose II intermolecular OH hydrogen bonds [25, 27, 28]
3230–3310	Cellulose I O(6)H–O(3) hydrogen bonding [25, 27, 28]
3200–3400	[O—H] stretching in strongly hydrogen bonded hydroxyls [20–22, 24]
3175	Cellulose II [O—H] stretching [25, 27, 28]
3100–3400	[O—H] stretching of hydroxyl groups in coir fibre hemicelluloses absent or significantly reduced after mercerisation [33]
3000–2800	[O—H] band absent or significantly reduced after mercerisation [33]
2925	Untreated sisal fibres, aliphatic [O—H] stretch [42]
2925	Acetylated sisal fibres, aliphatic [C—H] stretch [42]
2840–2940	[C—H] stretching [20–22, 24]
1740	Associated with [C=O] stretching vibrations from hemicellulose component of kapok fibre, absent after mercerisation [30] and xylan component of hemicellulose of flax and hemp fibres [36]
1739	Acetylated sisal fibres, carbonyl [C=O] stretch [42]
1738	Untreated jute and bamboo fibres acetyl and uronic ester groups or the ester linkage of carboxylic group of the ferulic and p-coumeric acids of hemicelluloses, absent in the cellulose extracted from these fibres after dewaxing, delignification and mercerisation [35]
1737	Untreated hemp fibres, [C=O] stretching in hemicellulose [41]
1735	Untreated sisal fibres, carbonyl [C=O] stretch of carboxyl and ester [42]
1732	Acetylated sunhemp fibres, carbonyl [C=O] stretch [42]
1726	Acetylated hemp fibres, [C=O] stretching in hemicellulose [41]
1723	Associated with hemicellulose component of coir fibre ([C=O]), absent after mercerisation [33]
1720–1740	[C=O] stretching in carbonyls [20–22, 24]
1650	NaOH treated hemp fibres, [C=O] stretching in hemicellulose [41]
1643	Untreated hemp fibres, [C=O] stretching in hemicellulose [41]
1643	Acetylated hemp fibres, [C=O] stretching in hemicellulose [41]

(*Continued*)

Table 2.4. (*Continued*)

Wavenumber (cm^{-1})	Association
1630	[–C=O] stretching (kapok fibre) [32]
1624	[–C=O] stretching (kapok fibre) after mercerisation [30]
1591	Untreated coir fibres [C=C] stretching of aromatic rings in lignin, absent in mercerised coir fibre [33]
1507	Untreated southern cattail fibres [C=O] stretching of amide groups in hemicelluloses and lignin, absent in mercerised southern cattail fibre [34]
1506	[C=C] stretching in aromatics [23]
1428	Untreated sisal fibres, [CH_3] bending [42]
1411	Acetylated sunhemp fibres, [CH_3] bending [42]
1402	Acetylated sisal fibres, [CH_3] bending [42]
1400–1430	[C–H] deformation due to methoxylate group in lignins [20]
1361	Untreated southern cattail fibres [C=O] stretching of amide groups in hemicelluloses and lignin, absent in mercerised southern cattail fibre [34]
1327	[CH_2] wagging in cellulose [20]
1327	[C–O] stretching in lignin (syringyl ring) [20–22, 24]
1293	Untreated coir fibres [C–O] stretching of the acetyl group of lignin, absent/ reduced in mercerised coir fibre [33]
1256	NaOH hemp fibres, [C=O] stretching in lignin [41]
1253	Acetylated hemp fibres, [C=O] stretching in lignin [41]
1250–1260	[C–O] stretching in lignin (guaiacyl ring) [20–22, 24]
1249	Untreated hemp fibres, [C=O] stretching in lignin [41]
1246	[C=O] and G ring stretching [23]
1244	Acetylated coir fibre [C–O] stretching of acetyl groups from acetylation [39]
1242.4	Untreated jute and bamboo fibres stretching vibrations of phenolic hydroxyl groups in lignin, absent in the cellulose extracted from these fibres after dewaxing, delignification and mercerisation [35]
1240	Untreated kapok fibre [C–O] aryl group lignin, absent after mercerisation [30, 31]
1240–1245	[C–O] bond stretching from acetyl groups in xylan and hemicelluloses [20]
1230	[O–H] deformation due to syringyl structure [23]
1229	Acetlyated flax and hemp fibres ([C–O]) stretching from acetates [36]
1162/1163	Acetylated flax and hemp fibres ([C–O]) stretching from ester carboxyl groups [36]
1150–1160	[C–O–C] antisymmetrical stretching in cellulose and aromatic [C–HCH_2] wagging in cellulose [20]
1098–1120	[C–O] stretching of p glycosidic linkages in cellulose [20]
1047	Acetylated sisal fibres, [C–O–C] [42]
1036	Acetylated flax and hemp fibres [C–O] stretching [20]
1046	Untreated sisal fibres, [C–O–C] [42]
1024	Acetylated sunhemp fibres, [C–O–C] [42]
1003	[C–O] stretching in cellulose [20]
890–900	Antisymmetric stretching of β linkages in cellulose [20]
830	[C–H] out of plane vibrations in lignin [20]

References

[1] Touaiti F, Pahlevan M, Nilsson R, Alam P, Toivakka M, Ansell M P and Wilen C E 2013 Impact of functionalised dispersing agents on the mechanical and viscoelastic properties of pigment coating *Progr. Org. Coat.* **76** 101–6

[2] Havigh R S and Chenari H M 2022 A comprehensive study on the effect of carbonization temperature on the physical and chemical properties of carbon fibers *Sci. Rep.* **12** 10704

[3] Alam P 2021 *Composites Engineering: An A-Z Guide* (Bristol: IOP Publishing)

[4] Brocks T, Cioffi M O H and Voorwald H J C 2013 Effect of fiber surface on flexural strength in carbon fabric reinforced epoxy composites *Appl. Surf. Sci.* **274** 210–6

[5] Dai Z, Shi F, Zhang B, Li M and Zhang Z 2011 Effect of sizing on carbon fiber surface properties and fibers/epoxy interfacial adhesion *Appl. Surf. Sci.* **257** 6980–5

[6] Ma J, Jiang L, Dan Y and Huang Y 2022 Study on the inter-laminar shear properties of carbon fiber reinforced epoxy composite materials with different interface structures *Mater. Des.* **214** 110417

[7] Zusron M *et al* 2016 Glass coating natural fibres by diatomisation: a bright future for biofouling technology *Mater. Today Commun.* **7** 81–8

[8] Jeevi G, Ranganathan N and Abdul Kader M 2022 Studies on mechanical and fracture properties of basalt/E-glass fiber reinforced vinyl ester hybrid composites *Polym. Compos.* **43** 3609–25

[9] Ramlan N, Zubairi S I and Maskat M Y 2022 Response surface optimisation of polydimethylsiloxane (PDMS) on borosilicate glass and stainless steel (SS316) to increase hydrophobicity *Molecules* **27** 3388

[10] Prakash V R A and Rajadurai A 2017 Inter laminar shear strength behavior of acid, base and silane treated E-glass fibre epoxy resin composites on drilling process *Defence Technol.* **13** 40–6

[11] Xu K, Ou Y, Li Y, Su L, Lin M, Li Y, Cui J and Liu D 2020 Preparation of robust aramid composite papers exhibiting water resistance by partial dissolution/regeneration welding *Mater. Des.* **187** 108404

[12] Shebanov S M, Novikov I K, Pavlikov A V and Ananin O B 2016 IR and Raman spectra of modern aramid fibers *Fibre Chem.* **48** 158–64

[13] Derombise G, Chailleux E, Forest B, Riou L, Lacotte N, Vouyovitch Van Schoors L and Davies P 2011 Long-term mechanical behavior of aramid fibers in seawater *Polym. Eng. Sci.* **51** 1366–75

[14] Derombise G, Vouyovitch Van Schoors L and Davies P 2009 Degradation of Technora aramid fibres in alkaline and neutral environments *Polym. Degrad. Stab.* **94** 1615–20

[15] Dixit P, Ghosh A and Majumdar A 2019 Hybrid approach for augmenting the impact resistance of p-aramid fabrics: grafting of ZnO nanorods and impregnation of shear thickening fluid *J. Mater. Sci.* **54** 13106–17

[16] Chen J, Zhu Y, Ni Q, Fu Y and Fu X 2014 Surface modification and characterization of aramid fibers with hybrid coating *Appl. Surf. Sci.* **321** 103–8

[17] Li Y, Luo Z, Yang L, Li X and Xiang K 2019 Study on surface properties of aramid fiber modified in supercritical carbon dioxide by glycidyl-POSS *Polymers* **11** 700

[18] Jia C, Yuan C, Ma Z, Du Y, Liu L and Huang Y 2019 Improving the mechanical and surface properties of aramid fiber by grafting with 1,4-dichlorobutane under supercritical carbon dioxide *Materials* **12** 3766

[19] Elfaleh I, Abbassi F, Bahibi M, Ahmad F, Guedri M, Nasri M and Garnier C 2023 A comprehensive review of natural fibers and their composites: an eco-friendly alternative to conventional materials *Results Eng.* **19** 101271

[20] Mosiewicki M A, Marcovich N E and Aranguren M I 2011 Characterization of fiber surface treatments in natural fiber composites by infrared and Raman spectroscopy *Interface Engineering of Natural Fibre Composites for Maximum Performance* Woodhead Publishing Series in Composites Science and Engineering (Cambridge: Woodhead Publishing) ch 4 pp 117–45

[21] Marcovich N E, Reboredo M M and Aranguren M I 1996 FTIR spectroscopy applied to woodflour *Compos. Interfaces* **4** 119–32

[22] Bessadok A, Langevin D, Gouanve F, Chappey C, Roudesli S and Marais S 2009 Study of water sorption on modified agave fibres *Carbohydrate Polym.* **76** 74–85

[23] Dai D and Fan M 2010 Characteristic and performance of elementary hemp fibre *Mater. Sci. Appl.* **1** 336–42

[24] Jayaramudu J, Jeevan Prasad Reddy D, Guduril B R and Varada Rajulu A 2009 Properties of natural fabric polyalthia cerasoides *Fibers Polym.* **10** 338–42

[25] Fan M, Dai D and Huang B 2012 *Fourier transform infrared spectroscopy for natural fibres Fourier Transform, Materials Analysis ed S Salih* (Rijeke, Croatia and Shanghai, China: InTech) 45–68 pp ch 3

[26] Geminiani L, Campione F P, Corti C, Luraschi M, Motella S, Recchia S and Rampazzi L 2022 Differentiating between natural and modified cellulosic fibres using ATR-FTIR spectroscopy *Heritage* **5** 4114–39

[27] Kolpak F J and Blackwell J 1976 Determination of the structure of cellulose II *Macromolecules* **9** 273–8

[28] Kondo T 2005 Hydrogen bonds in cellulose and cellulose derivatives *Polysaccharides: Structural Diversity and Functional Versatility* ed S Dumitriu (Boca Raton, FL: CRC Press)

[29] Idicula M, Boudenne A, Umadevi L, Ibos L, Candau Y and Thomas S 2006 Thermophysical properties of natural fibre reinforced polyester composites *Compos. Sci. Technol.* **66** 2719–25

[30] Dimple S G P and Shekhawat M S 2023 The effect of mercerization on the physical, mechanical, morphological, and chemical properties of Aerva Tomentosa fiber and fiber-reinforced urea-formaldehyde composites *Mater. Today Proc.* **92** 522–9

[31] Troedec M L, Peyratout C, Chotard T, Bonnet J P, Smith A and Guinebretiere R 2007 Physico-chemical modifications of the interactions between hemp fibres and a lime mineral matrix: impacts on mechanical properties of mortars *10th Int. Conf. of the European Ceramic Society (Berlin, Germany, 17–20 June 2007)* J G Heinrich and G Aneziris pp 451–6

[32] Malenab R A J, Ngo J P S and Promentilla M A B 2017 Chemical treatment of waste abaca for natural fiber-reinforced geopolymer composite *Materials* **10** 579

[33] Dugvekar M and Dixit S 2023 Thermal, structural, and morphological examination of mercerized coir fiber-reinforced composites *Proc. Inst. Mech. Eng.* **27** 982–8

[34] Ramesh M, Deepa C, Tamil Selvan M and Hemachandra Reddy K 2022 Effect of alkalization on characterization of Ripe Bulrush (Typha Domingensis) grass fiber reinforced epoxy composites *J. Nat. Fibers* **19** 931–42

[35] Liew F K, Hamdan S, Rahman M R and Rusop M 2017 Thermomechanical properties of jute/bamboo cellulose composite and its hybrid composites: the effects of treatment and fiber loading *Adv. Mater. Sci. Eng.* **2017** 8630749

[36] Tserki V, Zafeiropoulos N E, Simon F and Panayiotou C 2005 A study of the effect of acetylation and propionylation surface treatments on natural fibres *Composites A* **36** 1110–8
[37] Celino A, Freour S, Jacquemin F and Casari P 2013 Characterization and modeling of the moisture diffusion behavior of natural fibers *J. Appl. Polym. Sci.* **130** 297–306
[38] Garside P and Wyeth P 2002 Identification of cellulosic fibres by FTIR spectroscopy *Stud. Conservation* **48** 269–75
[39] Khalil H P S A, Ismail H, Rozman H D and Ahmad M N 2001 The effect of acetylation on interfacial shear strength between plant fibres and various matrices *Eur. Polym. J.* **37** 1037–45
[40] Rana A K, Basak R K, Mitra B C, Lawther M and Banerjee A N 1997 Studies of acetylation of jute using simplified procedure and its characterization *J. Appl. Polym. Sci.* **64** 1517–23
[41] Wang H, Kabir M M and Lau K T 2014 Hemp reinforced composites with alkalization and acetylation fibre treatments *Polymers Polym. Compos* **22** 247–52
[42] Chand N, Verma S and Khazanchi A C 1989 SEM and strength characteristics of acetylated sisal fibre *J. Mater. Sci. Lett.* **8** 13307–9

IOP Publishing

Composite Interfaces in Mechanical Design

Parvez Alam

Chapter 3

Bonding mechanisms at composite interfaces

3.1 Introduction

The effectiveness of a composite in terms of its properties, behaviour, and performance is explicitly related to the effectiveness of bonding at its component interfaces. In turn, the strength and effectiveness of a bonded interface depends on one or more mechanisms of adhesion. The most commonly cited mechanisms include mechanical interlocking of adhesion, electrostatic adhesion, chemical adhesion, and diffusion adhesion. It should be noted that adhesion based on a single one of the mechanisms is atypical; rather, adhesion between two composite components will include more than one of the mechanisms, which are in some situations interdependent on one another. This chapter focusses on describing the different mechanisms of adhesion in the context of engineered composites.

3.2 Mechanical interlocking of adhesion

As will be elucidated in more detail in chapter 6, the surfaces of reinforcing fibres are never really 'flat'. Reinforcement surfaces will exhibit topographical irregularities, which are often defined as variations in profile from an idealised flat line or plane. When a reinforcement surface is sufficiently wet by an adhesive or by liquid matrix material, the topographical irregularities form physical blockages to shearing motion at the interface. The irregularities in effect create small surface regions that prevent local matrix movement at the interface, thereby increasing the mechanical energy required to enable shearing. As such, if we were to simplify topographical irregularities to geometrically simple peaks and troughs, figure 3.1, the effectiveness of mechanical interlocking could be understood as relying on (i) the wetting material remaining as a continuum of cohesively bonded matter **and** (ii) the continuum structure of the wetting material penetrating, at least to some extent, into the troughs.

Full wetting of material into the troughs of a topographically irregulary surface has the additional effect of increasing the surface area of adhesion between

doi:10.1088/978-0-7503-5688-6ch3

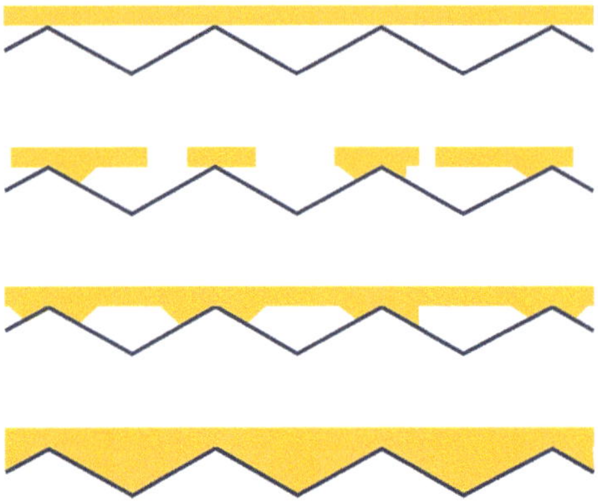

Figure 3.1. Depicting the effects of wetting material continuity, and wetting fraction, on the effectiveness of mechanical interlocking of adhesion.

reinforcement and matrix materials. If any other mechanisms of adhesion are concurrently taking place, these will naturally increase as a function of an increased surface area, again improving adhesion. A fully wet composite interface will also have an improved ability to retard fracture propagation as topographical irregularities are anfractuous and can either redirect or absorb energy at the crack front.

3.2.1 Fibre roughness

Roughness will differ based on the material used, the manufacturing method or processing conditions of the fibre, the fibre source, and on the surface treatment or finish. To detail such differences in roughness, we will provide examples using carbon fibres, table 3.1. In this table, roughness values are provided for a range of different carbon fibre types, processing parameters, and surface treatments. The initial data set (first band) represents data from Song and co-workers [1], who measured roughness using topographical image outputs by atomic force microscopy (scope: 3 μm × 3 μm). The carbon fibres used in their study were PAN-based T300 fibres (diameter 7 μm) manufactured by Jilin Chemical Industrial Company, China. These fibres were initially acetone treated to remove both impurities and sizing agent, then surface treated with aqueous ammonia (NH_3) and dried at 120 °C for 3 h, which results in both the erosion and oxidation of the carbon fibre surface [2]. The erosive nature of the NH_3 is evident as roughness can be seen to increase with prolonged exposure to NH_3.

The second band of data in table 3.1 is reported by Ruan and co-workers [3]. This data set is important as it informs us that similar fibres (polyacryonitrile-based 6K diameter 7 μm), each supplied by the same company (Weihai Guangwei Group Co. Ltd, China), can show significant variability in surface roughness. The fibres in this study were acetone treated and sonified to remove sizing.

The third band in table 3.1 shows a data set reported by Gao and co-workers [4], and considers the surface roughness of one intermediate modulus (IM) carbon fibre and two high modulus (HM) carbon fibres before and after single fibre pullout

Table 3.1. Example roughness values for a range of carbon fibres. IM = intermediate modulus, HM = high modulus, and PAN = polyacrylonitrile.

Roughness	Parameter	CARBON FIBRE Details
12.5 nm	Ra	PAN-based T300 (unsized) [1]
19.2 nm	Ra	PAN-based T300 (unsized) 24 h NH_3 treated [1]
25.2 nm	Ra	PAN-based T300 (unsized) 48 h NH_3 treated [1]
32.1 nm	Ra	PAN-based T300 (unsized) 72 h NH_3 treated [1]
42.3 nm	Ra	PAN-based T300 (unsized) 96 h NH_3 treated [1]
57.4 nm	Ra	PAN-based T300 (unsized) 120 h NH_3 treated [1]
1.6 nm	Ra	PAN-based 6 K type 1 (unsized) [3]
42 nm	Ra	PAN-based 6 K type 2 (unsized) [3]
78 nm	Ra	PAN-based 6 K type 3 (unsized) [3]
0.3 nm	Ra	IM (finished) [4]
4.2 nm ±2.5 nm	Ra	IM (finished) after pullout test from epoxy [4]
3.8 nm	Ra	HM (oxidised without finish) [4]
2.7 nm ±1.1 nm	Ra	HM (oxidised without finish) after pullout test from epoxy [4]
4.0 nm	Ra	HM (finished) [4]
3.7 nm ±1.3 nm	Ra	HM (finished) after pullout test from epoxy [4]
7.5 nm	Ra	PAN-based AS-4 (unsized) [5]
9.1 nm	RMS	PAN-based AS-4 (unsized) [5]
9.0 nm	Ra	PAN-based AS-4 (polyetherimide sized) [5]
11.0 nm	RMS	PAN-based AS-4 (polyetherimide sized) [5]
15.0 nm	Ra	PAN-based AS-4 (poly(thioarylene phosphine oxide) sized) [5]
18.0 nm	RMS	PAN-based AS-4 (poly(thioarylene phosphine oxide) sized) [5]
37 nm ± 17 nm	Ra	Panex 35 PAN-based 50 K tow automotive grade fibres (carbonised (unoxidised)) [6]
50 nm ± 10 nm	Ra	Panex 35 PAN-based 50 K tow automotive grade fibres (oxidised) [6]
33 nm ± 13 nm	Ra	Panex 35 PAN-based 50 K tow automotive grade fibres (sized) [6]
1.45 nm	Ra	Magnamite® IM6 (unsized) [7]
0.78 nm	Ra	Magnamite® IM6 (20% of nominal sizing) [7]
0.60 nm	Ra	Magnamite® IM6 (100% of nominal sizing) [7]
1.08 nm	Ra	Magnamite® IM6 (200% of nominal sizing) [7]
2.00 nm	Ra	Magnamite® IM6 (600% of nominal sizing) [7]

testing from an epoxy resin substrate. While the roughness of the IM fibre is a factor of 10 lower than those reported for HM fibres, this does not affect the roughness of the epoxy layer attached to the surface of the fibre after pullout testing, as deduced from the overlapping standard deviations between the three groups.

The fourth band of table 3.1 represents data sets reported by Dilsiz and Whiteman [5]. The data compares Hercules polyacrylonitrile-based carbon fibers (AS-4) using both Ra and RMS roughness. Here, the data highlights the importance of the roughness model used to determine roughness, with RMS roughness consistently providing higher nm values in this data set than the Ra roughness. Nevertheless, both roughness models are relatively consistent in terms of the differences detected, with polythioarylene phosphine oxide-sized fibres revealing the highest roughness, followed by polyetherimide-sized fibres, followed by unsized carbon fibres.

Possible differences in surface roughness at the different stages of manufacture leading to finished carbon fibres can be highlighted in the fifth band of table 3.1 using data published by Kafi and co-workers [6]. Here the authors compared the surface roughness of Panex 35 PAN-based 50 K tow automotive grade fibres supplied by Zoltek Companies, Inc., Hungary at the fibre processing stages of carbonisation, oxidation, and post-sizing (by epoxy). There are noticeably large differences in the mean from stage to stage, but the standard deviations overlap in all cases, elucidating the absence of tangible statistical evidence for differences in roughness data at each stage of manufacture.

The final band of table 3.1 shows data from Drzal and co-workers [7], and comparisons made between the surfaces of unsized Magnamite® IM6 fibres (0%) and Magnamite® IM6 fibres treated by an electrolytic anodisation process to 20%, 100%, 200%, and 600% of their nominal sizing level. The abrupt increase in roughness as a function of anodising beyond 100% is a result surface etching from the anodic process.

While the examples from each band in the list from table 3.1 are non-exhaustive, they represent a range of differences commonly observed in reinforcing fibres.

3.2.2 The effects of roughness on wetting

The effectiveness of mechanical interlocking is fundamentally governed by matrix wetting, which can be understood by measurement of the equilibrium contact angle, θ, of a droplet on a surface, as described by Young's equation, equation (3.1) [8], where γ represents the interfacial energy between the couplings of the three phases: solid (S), liquid (L), and gas (G). This equation ignores gravity as well as energy at the phase boundary. Furthermore, Gibbs [9] considered that residual energy would exist per unit length, as interface energies could not account for all free energies in the system. As such, a modification to Young's equation which accounts for excess energy (unaccounted for in equation (3.1)) is shown in equation (3.2), where a is the radius of the droplet and κ is the line tension.

$$\gamma_{SG} = \gamma_{SL} + \gamma_{LG} \cdot \cos\theta \tag{3.1}$$

$$\cos\theta = \frac{\gamma_{SG} - \gamma_{SL}}{\gamma_{LG}} + \frac{\kappa}{\gamma_{LG}}\frac{1}{a} \tag{3.2}$$

Reinforcement surfaces vary in roughness. As such the classical Young's equation used to determine the contact angle yields a typically inaccurate estimation of the contact angle, from which the effectiveness of wetting may be inferred. This is because Young's equation assumes an ideally flat surface, which is atypically observed on reinforcement surfaces. There are two levels at which modifications may be applied to Young's equation. The first is at a small enough length scale, where individual sufficiently small droplets wetting a rough surface form either within a concavity or upon a convexity of the topographically irregular profile. The second considers modifications at a macroscale, such that the global effect of roughness is built into contact angle estimations where Young's equation is deficient.

In the first case, where droplets are at a sufficiently small length scale to be modelled as wetting either the concavities or convexities of topographically irregular surfaces we can refer to the work of Jasper and Anand [10] who derived their contact angle models based on concave and convex structures with perfectly circular profiles, figure 3.2. Their model, equation (3.3), considers a roughness parameter, α, which is related to the curvature of the solid surface, and contains additional parameters, A, B, and C, represented by equations (3.4), (3.5), and (3.6), respectively, where A considers the bulk thermodynamic properties of wetting, B takes account of excess line energy at the interface triple-point (where the three phases meet), and in C the change in free energy of the liquid due to bulk changes in the curvature at the boundary due to a Laplace pressure, γ. In equation (3.3), the contact angle is $\cos(\theta + \alpha)$ in the case of concave structures and $\cos(\theta - \alpha)$ in the case of convex structures.

$$\cos(\theta \pm \alpha) = A + B\frac{\cos \alpha}{a} \pm C \sin(\theta \pm \alpha)(1 + \cos \theta)^2 \left(\frac{\sin \alpha(2 + \cos \alpha)}{(1 + \cos \alpha)^2} \pm \frac{\sin \theta(2 + \cos \theta)}{(1 + \cos \theta)^2} \right) \tag{3.3}$$

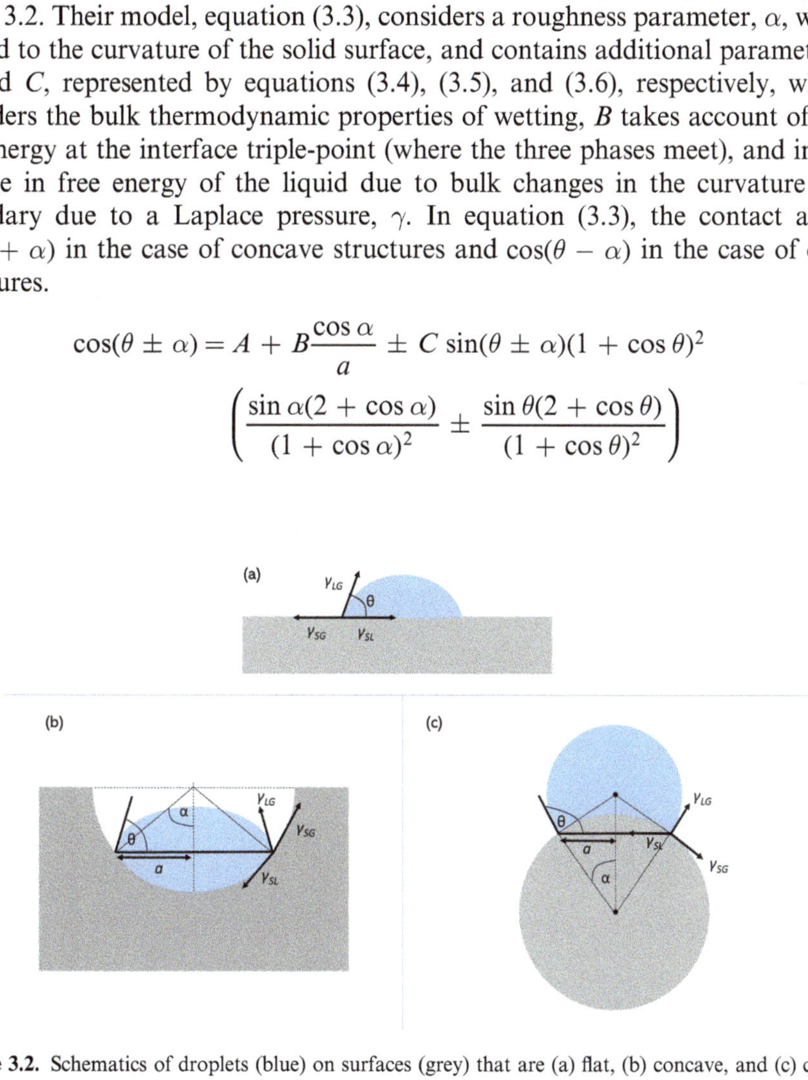

Figure 3.2. Schematics of droplets (blue) on surfaces (grey) that are (a) flat, (b) concave, and (c) convex.

$$A = \frac{\gamma_{SG} - \gamma_{SL}}{\gamma_{LG}} \tag{3.4}$$

$$B = \frac{\kappa}{\gamma_{LG}} \tag{3.5}$$

$$C = \frac{\gamma}{3\gamma_{LG}} \tag{3.6}$$

In the second case, where Young's equation is modified to incorporate surface roughness, a classic model is the Wenzel roughness model, r [11], which is shown in equation (3.7) and is essentially a ratio of the true surface area of the rough surface, A_r, and its nominal (perfectly flat) surface area, A_n, with the same external boundaries. The Wenzel contact angle, θ_w, is typically applied to Young's contact angle, as shown in equation (3.8). The Wenzel roughness predicts a perfectly flat surface as $r = 1$ and rough surfaces are $r > 1$, with the roughest surfaces showing $r \gg 1$.

$$r \frac{A_r}{A_n} \tag{3.7}$$

$$\cos\theta_w = r \cdot \cos\theta \tag{3.8}$$

Several modifications have been made to the Wenzel roughness to incorporate greater detail on roughness and its characteristics. One well known example is the Cassie–Baxter modification [12], which considers the fraction of the solid-to-liquid interface, f_{SL}, as well as the fraction of gaseous phase trapped within the troughs of a rough surface, $(1 - f_{SL})$, figure 3.3. The Cassie–Baxter contact angle, θ_{cb}, is shown in equation (3.9). If applying Cassie–Baxter to heterogenous and complex rough surfaces, the r is commonly included in the contact angle calculation as shown in equation (3.10) [13].

$$\cos\theta_{cb} = f_{SL}\cos\theta - (1 - f_{SL}) \tag{3.9}$$

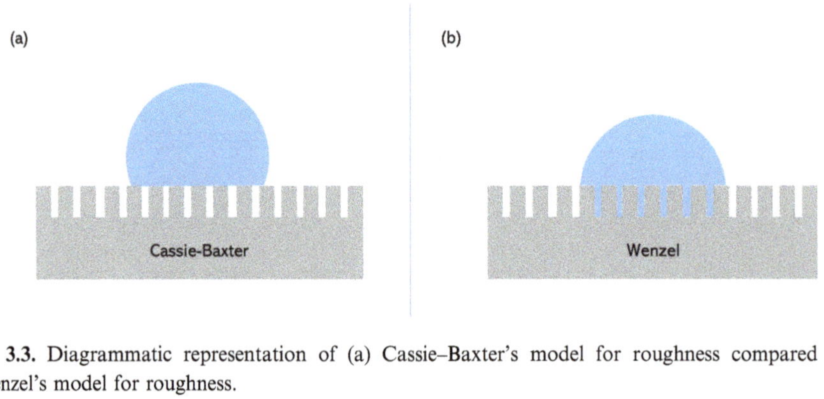

Figure 3.3. Diagrammatic representation of (a) Cassie–Baxter's model for roughness compared against (b) Wenzel's model for roughness.

$$\cos\theta_{cb} = r \cdot f_{SL}\cos\theta - (1 - f_{SL}) \tag{3.10}$$

Another modification to the Wenzel roughness derived by Alam and co-workers [14] considered the roughness, ϕ, as being defined by the product of the Wenzel roughness and an aspect ratio term, A_r, divided by the natural number raised to the power of a constant that considers the frequency of the protuberances as they appear on a rough surface. This is a semi-empirical roughness model shown in equation (3.11), where the aspect ratio term, A_r, is calculated according to equation (3.12). Here, $\bar{h}$ is the height average of the protuberances across the rough surface and $\overline{C_p}$ is the average cross-section of all protuberances. The exponent of the natural number in the denominator of equation (3.11) is calculated according to equation (3.13), where Δx represents a finite linear length across the which protuberance frequency is measured, $\overline{d_{pr}}$ represents the average protuberance edge-to-edge length along x, and n_0 is a unit scaling factor since the units of the edge and the protuberances may be several orders of magnitude different. The modified contact angle, θ_a, is shown in equation (3.14).

$$\phi = \frac{rA_r}{e^F} \tag{3.11}$$

$$A_r = \frac{\bar{h}}{\sqrt{\overline{C_p}}} \tag{3.12}$$

$$F = \frac{\Delta x}{\overline{d_{pr}}} \cdot n_0 \tag{3.13}$$

$$\cos\theta_a = \phi \cdot \cos\theta \tag{3.14}$$

The addition of nanoparticulates to reinforcement surfaces is becoming progressively more common as doing so can be useful in achieving specific goals such as increasing the effective surface area, developing surface hydrophobicity, altering the surface energy, and more. To this effect, the Wenzel model has received some treatment as it lacks applicability to nanostructured surfaces. A simple modification has been proposed by Dong and co-workers [15], which separates the surface into a fraction covered by, and not covered by, nanoparticulates, equation (3.15). In this equation, the contact angle, θ_d, takes into account the ordinary contact angle of the solid, θ, the relative area covered by nanoparticles, c, a shape factor for the protuberances, A, and the contact angle of the nanoparticulate surface, θ_{NP}.

$$\cos\theta_d = (1 - c)\cos\theta + Ac\cos\theta_{NP} \tag{3.15}$$

The combination of porosity and surface nanostructuring adds a further element of complexity to surface wetting and deductions of the equilibrium contact angle. In this light, Han and co-workers [16] proposed a modification to Wenzel's model for specifically hydrophobic behaviour that is observed on porous nanostructured surfaces. Their model, equation (3.16), calculates contact angle, θ_h, using the projected flat surface area, S, the total determined micropore volume, W_0, the pore size, X, expressed in terms of its distribution, the half width of micropores, X_0, and a dispersion parameter, δ.

$$\cos\theta_h = \frac{\cos\theta}{S}\int_{X_{min}}^{X_{max}} \frac{W_0}{\delta\sqrt{2\pi}}\frac{1}{X}\left[-\frac{(X_0 - X)^2}{2\delta^2}\right]\mathrm{d}X \tag{3.16}$$

3.3 Electrostatic adhesion

Electrostatic adhesion is a result of attraction between positively (cationic) and negatively (anionic) charged surfaces, figure 3.4. These charges may be distributed across a surface, or may be isolated to exposed charged functional groups of surface molecules. Essentially, interfacial strength is related to the charge densities [17]. One benefit of electrostatic adhesion is that unlike other secondary forces of attraction such as van der Waals, it is not as limited by surface-to-surface distance. While van der Waals forces may be limited to a few atomic diameters, the active range for electrostatic forces of attraction is in the centimetre range [18].

Electrostatic bonding has often been seen as a bonding mechanism confined primarily to metal-polymer interfaces [19, 20]; however, fibre surface treatments are modifiers of adhesion and are able to induce electrostatic effects between fibre–matrix materials that would not ordinarily be electrostatically compatible. Fibre sizing, coupling agents [21], and electrostatic discharge [22], for example, can be applied with the objective of increasing bond strength through improved electrostatic interactions [23], though the resultant strength improvement over parallel mechanisms of adhesion may be low. Small improvements through application of silane sizing have been shown to be possible on acidic or neutral reinforcement surfaces such as glass, silica, and aluminium oxide-based fibres. Silane sizing is nevertheless of less benefit when applied to alkaline surfaces, such as those found on magnesium and calcite [24]. The effectiveness of electrostatic forces can therefore also rise when certain types of sizing are removed from a reinforcement surface. For example, aramid fibres (Kevlar 49) can close to double their surface free energies from 33.3 to 54.1 mN m^{-1} when stripped of antistatic sizing [25]. Polyanaline coatings applied to carboxylated polyacrylonitrile fabric improve bond strength and bending stability through improved electrostatic adhesion [26], and the electrostatic adhesion of polymer particles to carbon fibre surfaces is also a method by which composites shear strength can be improved [27].

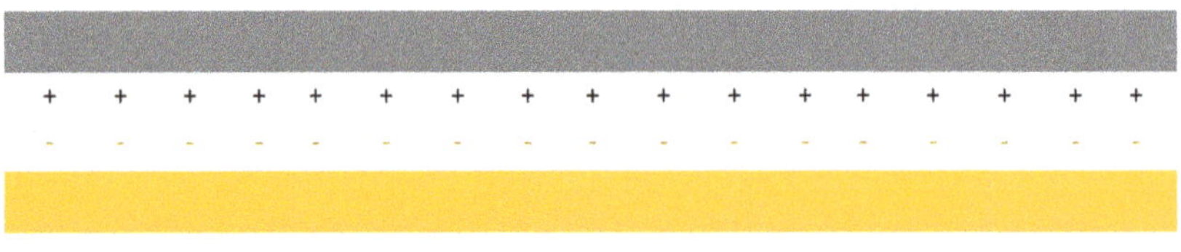

Figure 3.4. Diagrammatic representation of electrostatic adhesion between two oppositely charged surfaces.

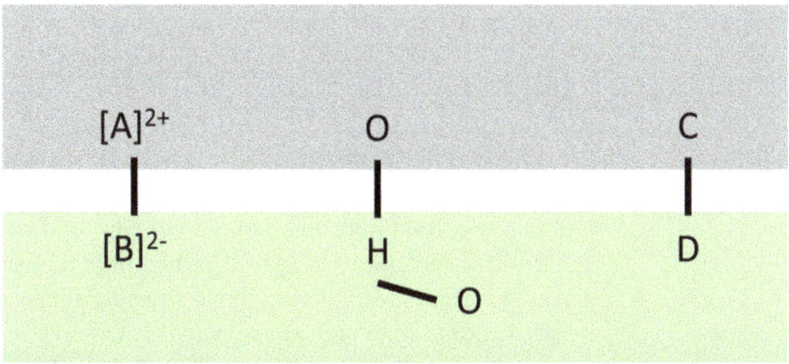

Figure 3.5. Diagrammatic representation of chemical bonding between two composite component surfaces showing ionic bonding (A–B), hydrogen bonding (H–OH), and covalent bonding (C–D).

3.4 Chemical adhesion

Chemical adhesion results in the formation of ionic, covalent, or hydrogen bonded interfaces between composite components. These are illustrated in figure 3.5, where ionic bonding is represented by A–B, hydrogen bonding by H—OH, and covalent bonding by C–D.

Many synthetic fibres such as aramids and glass exhibit typically inert surfaces and as such need to be functionalised by sizing or surface treatment. Glass fibres are a common inert engineering fibre and are most typically sized using silane coupling agents. Aramid fibres are often sized with polyurethane dispersions as these will react with resins during compounding processes [28]. There are three common types of carbon fibre modification: oxidative treatments, surface functionalisation using polymers, and particle modification to the surface, often with graphite, carbon powder, or other micro/nanoparticulates [29]. Natural fibres types are cellulose based and are subject to a wide range of fibre treatments including alkalisation, acetylation, polymer grafting, and others. Surface treatments are effectively enablers for chemical bonding at composite interfaces and are covered in more detail in chapter 5 of this book. A few examples are briefly discussed in table 3.2.

3.5 Diffusion adhesion and entanglements

Diffusion bonding relies on there being sufficient polymer mobility at interfaces between two surfaces. When two such surfaces make contact, polymer ends and portions of polymer molecules on at least one of the interfaces diffuse past the interface of the adjacent surface penetrating the bulk matter. This may occur bidirectionally, and essentially, polymer chain diffusion creates both loose physical molecular interlocking and associated intermolecular secondary force interactions, creating an interfacial bond. When molecular chain lengths from one material reach a critical length (or above) they may also physically entangle with molecules from the adjacent material they have diffused into. Figure 3.6 illustrates both cases of short chain molecular diffusion, and longer chain molecular entanglements that form during the process of diffusion bonding.

Table 3.2. Examples of chemical bonding at interfaces as applied to composite materials and their components.

Type	Details
Graphene oxide to epoxy	Hydroxyl and carboxyl groups of graphene oxide react with epoxy groups forming strong chemical bonds. The combination can be used to size carbon fibre and due to both improved compatibility, higher surface energies, changes to carbon fibre topography and increased surface area, interface bond strength is increased between the sized carbon fibre and an epoxy matrix [30, 31].
Polyurethane grafted carbon fibre	Elecrochemically oxidised carbon fibre grafted with hyperbranched polyurethane introduces amino groups that results in the covalent bonding of the grafted fibre to an epoxy matrix [32].
Polyacrylate sized carbon fibre	Polyacrylate emulsion used to size carbon fibres yielding improved interface adhesion through the creation of covalent bonds [33].
Silanised natural fibres	Silane coupling agents applied to natural fibres form silanols in the presence of moisture. The silanols react with hydroxyl groups on the fibre surface to create covalent bonds. The application of silane decreases swelling due to the presence of hydrocarbons from the coupling agent, and these cross-link covalently with the polymer matrix [34].
Acrylated natural fibres	Acrylic acid-modified natural fibre surfaces develop strong covalent bonds with matrix material. In particular, acrylated flax embedded into high-density polyethylene increases bond strength and decreases water ingress into the composites [35–37].
Maleated cellulose/polypropylene	Maleic anhydride bonds cohesively to polypropylene, which can be used to size cellulose-based reinforcing fibres to equalise the surface energies of the fibre with the matrix. The composite forms covalent bonds across the sizing-matrix interface, which stabilises the composite [38].
Plasma treated Twaron®/PPESK	Air dielectric barrier discharge (DBD) plasma surface-treated Twaron® aramid fibres used to reinforce thermoplastic poly (phthalazinone ether sulfone ketone), or PPESK. Interfacial adhesion improvements attributed to increased oxidation at the aramid fibre interface from plasma treatment (C=O), leading to improved chemical bonding of the fibres to PPSEK as well as increased opportunities for hydrogen bonding [39].
Epoxy-amine sized Twaron®	Twaron® aramid fibres surface activated using epoxy/amine sizing enabling their use as reinforcement in epoxy matrix composites [40].

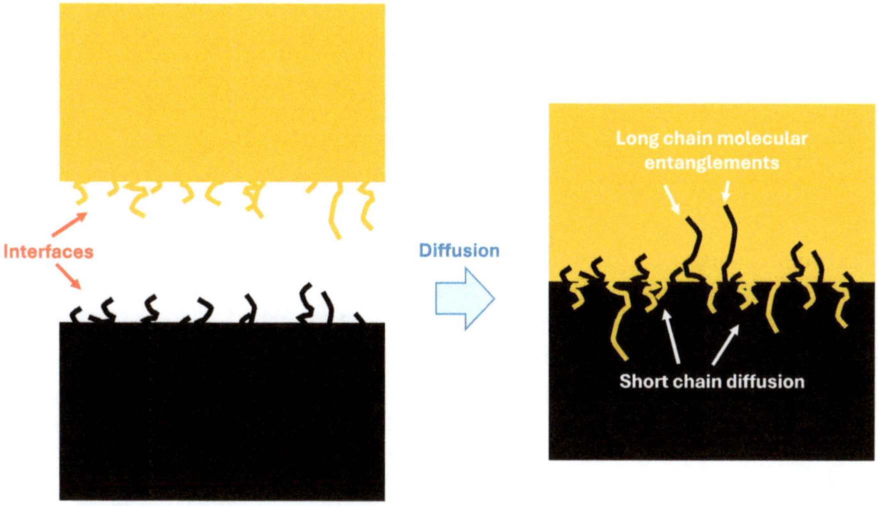

Figure 3.6. Schematic showing of short chain diffusion bonding and long chain molecular entanglements at the interfaces of joined polymers.

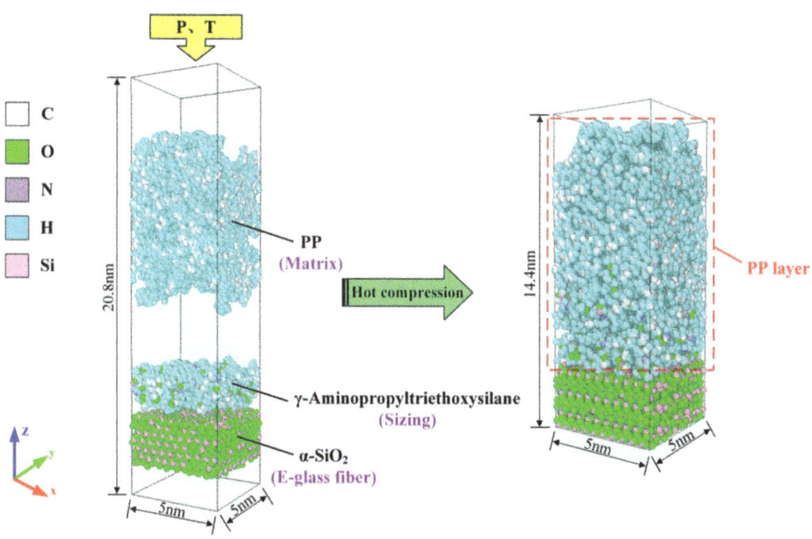

Figure 3.7. Molecular dynamics simulation showing diffusion bonding between γ-aminopropyltriethoxysilane (an E-glass sizing) and a polypropylene matrix. Reproduced from [41]. CC-BY 4.0.

Since diffusion bonding requires a certain level of molecular mobility at the interfaces of composites, it is an adhesion mechanism that is generally confined to matrix-sizing interfaces. This can be visualised in the molecular model shown in figure 3.7 from [41], where γ-aminopropyltriethoxysilane sizing applied to the surface of E-glass fibre is seen to penetrate several nanometres into polypropylene matrix when the composite components are hot compressed together.

References

[1] Song W, Gu A, Liang G and Yuan L 2011 Effect of the surface roughness on interfacial properties of carbon fibers reinforced epoxy resin composites *Appl. Surf. Sci.* **257** 4069–74

[2] Meng L H, Chen Z W, Song X L, Liang Y X, Huang Y D and Jiang Z X 2009 Influence of high temperature and pressure ammonia solution treatment on interfacial behavior of carbon fiber/epoxy resin composites *J. Appl. Polym. Sci.* **113** 3436–41

[3] Ruan R, Cao W and Xu L 2020 Quantitative characterization of physical structure on carbon fiber surface based on image technique *Mater. Des.* **185** 108225

[4] Gao S L, Mader E and Zhandarov S F 2004 Carbon fibers and composites with epoxy resins: topography, fractography and interphases *Carbon* **42** 515–29

[5] Dilsiz N and Wightman J P 1999 Surface analysis of unsized and sized carbon fibers *Carbon* **37** 1105–14

[6] Kafi A, Huson M, Creighton C, Khoo J, Mazzola L, Gengenback T, Jones F and Fox B 2014 Effect of surface functionality of PAN-based carbon fibres on the mechanical performance of carbon/epoxy composites *Compos. Sci. Technol.* **94** 89–95

[7] Drzal L T, Sugiura N and Hook D 1997 The role of chemical bonding and surface topography in adhesion between carbon fibers and epoxy matrices *Compos. Interfaces* **4** 337–54

[8] Young T 1805 An essay on the cohesion of fluids *Phil. Trans.* **95** 65

[9] Gibbs J W 1928 *The Collected Works of JW Gibbs* (New York: Longmans)

[10] Jasper W J and Anand N 2019 A generalized variational approach for predicting contact angles of sessile nano-droplets on both flat and curved surfaces *J. Mol. Liquids* **281** 196–203

[11] Wenzel R W 1936 Resistance of solid surfaces to wetting by water *Ind. Eng. Chem.* **28** 988–94

[12] Cassie A B D and Baxter S 1944 Wettability of porous surfaces *Trans. Faraday Soc.* **40** 546–51

[13] Kubiak K J, Wilson M C T, Mathia T G and Carval P H 2011 Wettability versus roughness of engineering surfaces *Wear* **271** 523–8

[14] Alam P, Toivakka M, Backfolk K and Sirvio P 2007 Impact spreading and absorption of Newtonian droplets on topographically irregular porous materials *Chem. Eng. Sci.* **62** 3142–58

[15] Dong L, Nypelo T, Osterberg M, Laine J and Alava M 2010 Modifying the wettability of surfaces by nanoparticles: experiments and modeling using the Wenzel law *Langmuir* **26** 14563–6

[16] Han T Y, Shr J F, Wu C F and Hsieh C T 2007 A modified Wenzel model for hydrophobic behavior of nanostructured surfaces *Thin Solid Films* **515** 4666–9

[17] Kim J K and Mai Y W 1998 Characterization of interfaces *Engineered Interfaces in Fiber Reinforced Composites* (Amsterdam: Elsevier Science) pp 5–41

[18] Mohammed M, Rasidi M S M, Mohammed A M, Rahman R, Osman A F, Adam T, Betar B O and Dahham O S 2022 Interfacial bonding mechanisms of natural fibre-matrix composites: an overview *Bioresources* **17** 7031–90

[19] Amiandamhen S O, Meincken M and Tyhoda L 2020 Natural fibre modification and its influence on fibre-matrix interfacial properties in biocomposite materials *Fibers Polym.* **21** 677–89

[20] Liu J, Xue Y, Dong X, Fan Y, Hao H and Wang X 2023 Review of the surface treatment process for the adhesive matrix of composite materials *Int. J. Adhes. Adhes.* **126** 103446

[21] Rao J, Zhou Y and Fan M 2018 Revealing the interface structure and bonding mechanism of coupling agent treated WPC *Polymers* **10** 266
[22] Rajak D K, Pagar D D, Menezes P L and Linul E 2019 Fibre-reinforced polymer composites: manufacturing, properties, and applications *Polymer* **11** 1667
[23] Arulvel S, Reddy D M, Rufuss D D W and Akinaga T 2021 A comprehensive review on mechanical and surface characteristics of composites reinforced with coated fibres *Surf. Interfaces* **27** 101449
[24] Teklal F, Djebbar A, Allaoui S, Hivet G, Joliff Y and Kacimi B 2018 A review of analytical models to describe pull-out behavior–fiber/matrix adhesion *Compos. Struct.* **201** 791–815
[25] Kalantar J and Drzal L T 1990 The bonding mechanism of aramid fibres to epoxy matrices. Part 1 A review of the literature *J. Mater. Sci.* **25** 4186–93
[26] Ke F, Zhang Q, Ji L, Zhang Y, Zhang C, Xu J, Wang H and Chen Y 2021 Electrostatic adhesion of polyaniline on carboxylated polyacrylonitrile fabric for high-performance wearable ammonia sensor *Compos. Commun.* **27** 100817
[27] Yamamoto T, Uematsu K, Irisawa T and Tanabe Y 2016 Controlling of the interfacial shear strength between thermoplastic resin and carbon fibre by adsorbing polymer particles on carbon fibre using electrophoresis *Composites A* **88** 75–8
[28] Michelman Hydrosize® 2018 Fiber sizings and composites. Enhancing composite performance through optimized fiber-polymer interface adhesion. 2018 Michelman, Edition: 03/2018 Global
[29] Banerjee P, Raj R, Kumar S and Bose S 2021 Tuneable chemistry at the interface and selfhealing towards improving structural properties of carbon fiber laminates: a critical review *Nanoscale Adv.* **3** 5745
[30] Pathak A K, Borah M, Gupta A, Yokozeki T and Dhakate S R 2016 Improved mechanical properties of carbon fiber/graphene oxide-epoxy hybrid composites *Compos. Sci. Technol.* **135** 28–38 Tuneable chemistry at the interface and selfhealing towards improving structural properties of carbon fiber laminates: a critical review
[31] Zhang X Q, Fan X Y, Yan C, Li H Z, Zhu Y D and Li X T 2012 Interfacial microstructure and properties of carbon fiber composites modified with graphene oxide *ACS Appl. Mater. Interfaces* **4** 1543–52
[32] Andideh M and Esfandeh M 2017 Effect of surface modification of electrochemically oxidized carbon fibers by grafting hydroxyl and amine functionalized hyperbranched polyurethanes on interlaminar shear strength of epoxy composites *Carbon* **123** 233–42
[33] Yuan X, Zhu B, Cai X, Liu J, Qiao K and Yu J 2017 Optimization of interfacial properties of carbon fiber/epoxy composites via a modified polyacrylate emulsion sizing *Appl. Surf. Sci.* **401** 414–23
[34] Agrawal R, Saxena N S, Sharma K B, Thomas S and Sreekala M S 2000 Activation energy and crystallization kinetics of untreated and treated oil palm fibre reinforced phenol formaldehyde composites *Mater. Sci. Eng.:* A **277** 77–82
[35] Sreekala M S, Kumaran M G, Joseph S, Jacob M and Thomas S 2000 Oil palm fibre reinforced phenol formaldehyde composites: influence of fibre surface modifications on the mechanical performance *Appl. Compos. Mater.* **7** 295–329
[36] Sreekala M S, Kumaran M G and Thomas S 2002 Water sorption in oil palm fiber reinforced phenol formaldehyde composites *Composites A* **33** 763–77
[37] Li X, Tabil L G and Panigrahi S 2007 Chemical treatments of natural fiber for use in natural fiber-reinforced composites: a review *J. Polym. Environ.* **15** 25–33

[38] Bledzki A K, Reihmane S and Gassan J 1996 Properties and modification methods for vegetable fibers for natural fiber composites *J. Appl. Polym. Sci.* **59** 1329–36

[39] Jia C, Chen P, Li B, Wang Q, Lu C and Yu Q 2010 Effects of Twaron fiber surface treatment by air dielectric barrier discharge plasma on the interfacial adhesion in fiber reinforced composites *Surf. Coat. Technol.* **204** 3668–75

[40] de Lange P J, Mader E, Kai K, Young R J and Ahmad I 2001 Characterization and micromechanical testing of the interphase of aramid-reinforced epoxy composites *Composites A* **32** 331–42

[41] Zhang M, Jiang B, Chen C, Drummer D and Zhai Z 2019 The effect of temperature and strain rate on the interfacial behavior of glass fiber reinforced polypropylene composites: a molecular dynamics study *Polymers* **11** 1766

IOP Publishing

Composite Interfaces in Mechanical Design

Parvez Alam

Chapter 4

From interfaces to interphases

4.1 Introduction

Composites are defined as materials comprising two or more materials with distinct interfaces between them. In polymer matrix composites, a direct consequence arising from the presence of an interface is the creation of an interphase. The interphase itself, its formation, its depth, and its properties are *at least* attributable to the type and rigidity of the polymer chains, their molecular weights, the strength of molecular pinning of polymer chains at the interface, the curing kinetics of the composite, the closeness of fibres within the matrix, and the interactions of polymer molecules with each other as they extend from the interface. As discussed by Drzal and co-workers [1], the interphase can be dimensionally defined as starting from where the fibre properties differ from the bulk fibre and ending at a point within the matrix where the properties of the interphase equal those of the bulk matrix.

When designing a composite, control over the fibre-spacing distances has the benefit of controlling the separation or coalescence of the interphase. Control of the spacing distances in turn is often most easily controlled by controlling external manufacturing variables such as consolidation pressures, control of curing kinetics, control of the fibre volume fraction, and the selection of specific matrix materials (e.g. curing rate, viscosity, Tg, etc.). The interphase may therefore exist in any of the following forms:

- a disconnected phase, i.e. interphases around fibres do not physically interact or share volume, figure 4.1(a);
- a partially connected phase, i.e. some fibre interphases are physically connected to other fibre interphases but the matrix bulk is still a predominantly continuous phase, figure 4.1(b); and
- a well-connected phase, i.e. many fibre interphases are physically connected to other fibre interphases becoming thence a continuous phase, leaving the matrix as a discontinuous phase in the composite, figure 4.1(c).

The interphase is essentially a 3D solid with properties and molecular conformations that are vastly different from both the 2D interface and the 3D bulk polymer.

doi:10.1088/978-0-7503-5688-6ch4

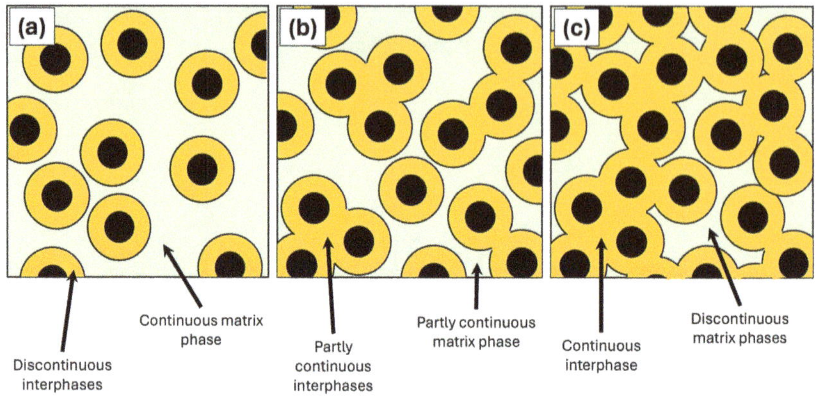

Figure 4.1. Examples of the effects of fibre-spacing distance on interphases in composites and how they affect the reinforcing and matrix phases: (a) fibre spacing is large enough to ensure that all interphase regions are separate and that the matrix remains a continuum material; (b) fibre spacing is close enough to create partially continuous interphases, but still far enough apart for the matrix phase to be a partly continuous phase; and (c) the fibre spacing is so close that the interphases form a continuous phase while the matrix phase exists as discontinuous phases within the composite.

In many fibre- reinforced composites, there is an additional complexity in defining the interface and the interphase, since fibres are sized and the sizing itself has its own properties, conformations, and layer thickness. Each of these factors influences the overall structure and thickness of the interphase. Whether the position of the 2D interface is at the fibre-sizing interface, at the sizing-matrix interface, or *somewhere in between* is a matter of philosophical debate and there is still no established position on the issue.

The interphase itself has been documented to have different total thicknesses, as evident from table 4.1. Jones [2] describes the development of structural differences, as determined using Time of Flight Secondary Ion Mass Spectrometry, between interphases in carbon fibre- reinforced composites using brominated epoxy-sized carbon fibres as a constant, with either thermoplastic or epoxy as variable matrices. Here, the interphase of the thermoplastic matrix composite is reported to develop in a graded manner, whereas in the epoxy matrix composite is reported as being a distinct single interphase. The structural differences indubitably have an effect on properties such as load bearing, resistance to shear, fracture profile, and more. In addition, the curing cycle has a notable effect on the final thickness of the interphase. By example, the average thickness of interphase in a fast cured composite has been shown to be half the thickness of interphase in a slow cured composite [3]. The logic here is that a slow cured interphase has more time to grow from sites of interfacial nucleation.

4.2 Models of interphase formation

Characterising interphase formation is complex as the phenomenon is influenced by a variety of factors including molecular mobility, molecular weight, molecular conformation, glass transition temperature, curing characteristics, the type and

Table 4.1. Collection of interphase thicknesses for different composite materials as reported in the literature.

Details	Interphase thickness	Source
Unidirectional carbon fibre fabrics T700SC-12000-50C (Toray Inc.) with epoxy (bisphenol A) Araldite LY 1564 with fast curing agent diethylenetriamine, DETA Hardener XB 3458 (Huntsman Inc.) 80 °C/10 min plus 100 ° C/20 min (fast cure)—determined using Peak Force Quantitative Nano-Mechanics tests using modulus image	20.03 ± 2.04 nm	Qi *et al* [3]
Unidirectional carbon fibre fabrics T700SC-12000-50C (Toray Inc.) with epoxy (bisphenol A) Araldite LY 1564 with fast curing agent diethylenetriamine, DETA Aradur R 3486 (Huntsman Inc.) 80 ° C/8 h (conventional cure)—determined using Peak Force Quantitative Nano-Mechanics tests using modulus image	40.48 ± 4.17 nm	Qi *et al* [3]
Unidirectional carbon fibre fabrics T700SC-12000-50C (Toray Inc.) with epoxy (bisphenol A) Araldite LY 1564 with fast curing agent diethylenetriamine, DETA Hardener XB 3458 (Huntsman Inc.) 80 ° C/10 min plus 100 ° C/20 min (fast cure)—determined using Peak Force Quantitative Nano-Mechanics tests based on adhesion contrast	19.45 ± 0.68 nm	Qi *et al* [3]
Unidirectional carbon fibre fabrics T700SC-12000-50C (Toray Inc.) with epoxy (bisphenol A) Araldite LY 1564 with fast curing agent diethylenetriamine, DETA Aradur R 3486 (Huntsman Inc.) 80 ° C/8 h (conventional cure)—determined using Peak Force Quantitative Nano-Mechanics tests based on adhesion contrast	41.01 ± 3.98 nm	Qi *et al* [3]
Optical grade silica fibre (Dolan-Jenner), difunctional bisphenol A/epichlorohydrin-derived liquid epoxy resin EPON 828 (Shell Chemical), Nadic methylanhydride (NMA) curing agent (Aldrich Chemical), γ-Aminopropyltriethoxysilane sizing—nanoindentation and atomic force microscopy	2.4–2.9 μm	Downing *et al* [4]
Carbon fibre/polyphenylenesulfide (PPS) prepreg (HM-type carbon fibre (Toray Inc.), PPS (Fortron, Hoechst))— scanning force microscopy	20–80 nm	Munz *et al* [5]

(*Continued*)

Table 4.1. (*Continued*)

Details	Interphase thickness	Source
Silica filler (Rockwood Clay Additives Inc.) reinforced polypropylene (TVK, Hungary) average molecular mass 97 kg mol^{-1}, polydispersity 5.0—model- based estimation based on experimentally determined mechanical properties	0.23 μm	Dominkovics *et al* [6]
Glass fibre-reinforced epoxy resin—atomic force microscopy in force modulation	1–3 μm	Mai *et al* [7]
E-glass roving (Owens Corning Reinforcements), pre-polymer of liquid DGEBA type and an aliphatic amine hardener (diethylenetriamine—DETA with purity at 99%), γ-AminoPropyltriethoxySilane sizing— identified by atomic force microscopy	500 nm	Riano *et al* [8]
T300 carbon fibre (Toray Inc.)/epoxy resin—characterisation by dynamic mechanical imaging	118 nm	Gu *et al* [9]
T700 carbon fibre (Toray Inc.)/bismaleimide resin—characterisation by dynamic mechanical imaging	163 nm	Gu *et al* [9]

depth of fibre sizing, and more. Pitchumani [10] discusses the formation of interphases in thermosetting polymer matrix composites both in terms of kinetics [11] and in terms of thermodynamics of both non-reacting [12] and reacting [13] fibre–matrix interfaces.

4.2.1 Kinetics of formation

Reinforcing fibres are known to affect the characteristics of curing as specific chemical species are adsorbed to fibre surfaces, which results in differential concentrations of reaction species near fibre surfaces. The kinetics model proposed by [11] uses processes that occur during curing, to predict the concentration profiles of constituent chemical species near the reinforcement surface. The cure kinetics model is based on the fibre arrangement shown in figure 4.2, where d is the diameter, R is the length between fibre centres (from cross- sections), and L is the distance between fibre surfaces. The model geometry as can be seen in figure 4.2(a) is in a staggered arrangement, while the idealised modelled region indicated by the dotted line in figure 4.2(b) is the centre plane between the two fibre surfaces.

There are three processes involved in the formation of the interphase: adsorption, desorption, and diffusion. In this model, the fibre surface is assumed to be surface sized by epoxy that is itself exposed to an epoxy–amine mixture. The epoxy, E, and

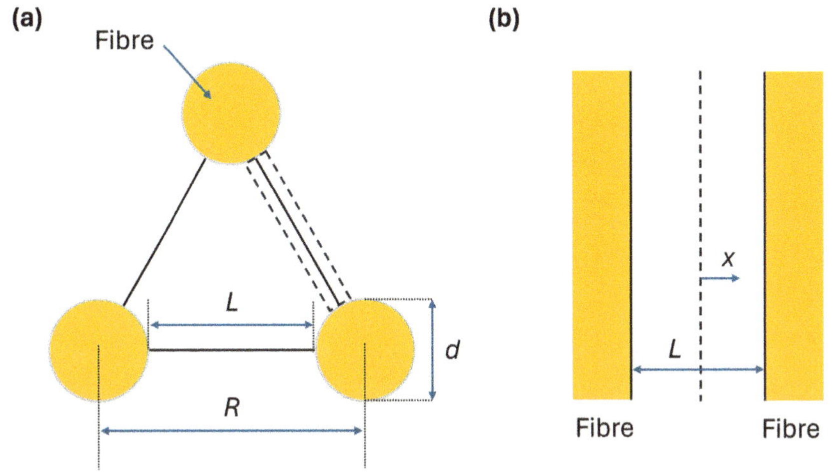

Figure 4.2. (a) Fibre arrangement in a composite relevant to the kinetics model of [11] and (b) one-dimensional domain between two fibres (their surfaces) considered in the kinetics model. Here d is the diameter, R is the length between fibre centres (from cross-sections), and L is the distance between fibre surfaces. In (b) the dotted line represents the model domain, i.e. the middle plane.

amine, A, react chemically during the adsorption and desorption processes, but a reaction term is included since reacting species will result in the depletion of molecules. The reaction is simply expressed by equation (4.1), where P is the product, while n_1 and n_2 are the molar numbers of reactants per 1 mol product. The model considers E and A spatial positions as being relative to the fibre surface, and each of E and A are at any ith layer from the surface of the fibre. Using the principle of mass conservation within a controlled volume of material, the rate of change $\left(\frac{dN_{Y,i}}{dt}\right)$ understood according to equation (4.2), where Y is equivalent to E if calculating the rate of change for epoxy species $\left(\frac{dN_{E,i}}{dt}\right)$, or is equivalent to A if calculating the rate of change for amine species $\left(\frac{dN_{A,i}}{dt}\right)$, t is time, and i refers to the ith layer to which the calculation refers. In equation (4.2), $\sum R^i_{a,Y}$ refers to the total adsorption in the ith layer and the two layers adjacent to it, $\sum R^i_{d,Y}$ refers to the total desorption in the ith layer and the two layers adjacent to it, and $R^i_{r,Y}$ refers to the reaction in the ith layer, and these are calculated in accordance with equations (4.3), (4.4), and (4.5), respectively. While both adsorption and desorption involve molecules moving from/to neighbouring layers, reaction occurs within a layer and as such the reaction rate is only considered for the layer itself, whereas adsorption and desorption take into consideration the neighbouring layers.

$$n_1E + n_2A \rightarrow P \tag{4.1}$$

$$\frac{dN_{Y,i}}{dt} = \sum R^i_{a,Y} - \sum R^i_{d,Y} - R^i_{r,Y} \tag{4.2}$$

$$\sum R^{i}_{a,Y} = R_{a,Y}(i-1, i) + R_{a,Y}(i, i) + R_{a,Y}(i+1, i) \tag{4.3}$$

$$\sum R^{i}_{d,Y} = R_{d,Y}(i-1, i) + R_{d,Y}(i, i) + R_{d,Y}(i+1, i) \tag{4.4}$$

$$R^{i}_{r,Y} = R_{r,Y}(i) \tag{4.5}$$

The individual reaction terms for adsorption, equation (4.3), are calculated according to equations (4.6) to (4.8), in respective order. Here, $k_{a,Y}$ is a frequency factor in the adsorption rate of molecules, if epoxy then Y is replaced by E and if amine then Y is replaced by A; the total number of molecules of E and A adsorbed in the ith layer is represented by N_i, such that $N_i = N_{E,i} + N_{A,i}$; and the term $N_{i-1} - N_i$ represents the number of available sites for adsorption in the $(i-1)$ th layer, given that an adsorbed molecule in the ith layer will adjoin to an epoxy or amine molecule in the $(i-1)$ th layer. The term $E_{a,Y}$ denotes the activation energy of adsorption of specific molecules; if epoxy, then $E_{a,E}$, and if amine then $E_{a,A}$. R is the universal gas constant and T is the temperature. $N_{Y\infty,i}$ represents either $N_{E\infty,i}$ or $N_{A\infty,i}$, which are the number of epoxy or amine molecules in their bulk state in the ith layer such that $N_{\infty,i} = N_{E\infty,i} + N_{A\infty,i}$. The fraction $\frac{N_{Y\infty,i-1}}{N_{\infty,i-1}+N_{\infty,i}+N_{\infty,i+1}}$ is the probability a site can capture a molecule from the $(i-1)$ th layer of the bulk state, whether it be for epoxy, $\frac{N_{E\infty,i-1}}{N_{\infty,i-1}+N_{\infty,i}+N_{\infty,i+1}}$ or for amine $\frac{N_{A\infty,i-1}}{N_{\infty,i-1}+N_{\infty,i}+N_{\infty,i+1}}$.

$$R_{a,Y}(i-1, i) = k_{a,Y}(N_{i-1} - N_i) \times \exp\left(-\frac{E_{a,Y}}{RT}\right)\frac{N_{Y\infty,i-1}}{N_{\infty,i-1} + N_{\infty,i} + N_{\infty,i+1}} \tag{4.6}$$

$$R_{a,Y}(i, i) = k_{a,Y}(N_{i-1} - N_i) \times \exp\left(-\frac{E_{a,Y}}{RT}\right)\frac{N_{Y\infty,i}}{N_{\infty,i-1} + N_{\infty,i} + N_{\infty,i+1}} \tag{4.7}$$

$$R_{a,Y}(i+1, i) = k_{a,Y}(N_{i-1} - N_i) \times \exp\left(-\frac{E_{a,Y}}{RT}\right)\frac{N_{Y\infty,i+1}}{N_{\infty,i-1} + N_{\infty,i} + N_{\infty,i+1}} \tag{4.8}$$

When defining the general rate equations for desorption, $R_{d,Y}$, Pitchumani considers that $R_{d,Y}(i-1, i) = R_{d,Y}(i, i) = R_{d,Y}(i+1, i)$ and each is therefore represented by equation (4.9). Here, the fraction $\frac{1}{3}$ assumes the probabilities of desorption from the ith layer to the neighbouring layers are identical. $k_{d,Y}$ is a frequency factor in the desorption rate of molecules; if epoxy, then Y is replaced by E and if amine then Y is replaced by A. Similarly, the term $E_{d,Y}$ denotes the activation energy of desorption of specific molecules; if epoxy, then $E_{d,E}$, and if amine then $E_{d,A}$. Finally, $N_{Y,i}$ represents the number of molecules of Y (either E or A) in the ith layer.

$$\begin{aligned} R_{d,Y}(i-1, i) &= R_{d,Y}(i, i) = R_{d,Y}(i+1, i) \\ &= \frac{1}{3}k_{d,Y}(N_i - N_{i+1}) \times \exp\left(-\frac{E_{d,Y}}{RT}\right)\frac{N_{Y,i}}{N_i} \end{aligned} \tag{4.9}$$

The general rate equation for reacting species is shown in equation (4.10), where $k_{r,Y}$ is a frequency factor in the reaction rate of molecules; if epoxy, then Y is replaced by E and if amine then Y is replaced by A.

$$R_{r,Y}(i) = n_1 k_r N_{Y,i} \tag{4.10}$$

There is an exchange of mass between the adsorbed and bulk states where molecules will diffuse alongside the ongoing adsorption, desorption, and reaction processes. These are broken down into the following four processes:

1. The diffusion of epoxy molecules in the bulk from $(i + 1)$th and $(i - 1)$th layers to the ith layer, which results in an increase in $N_{E\infty,i}$.
2. The desorption of epoxy molecules in the adsorbed state from the $(i + 1)$ th, ith, and $(i - 1)$th layers to the ith layer of the bulk, which results in an increase in $N_{E\infty,i}$.
3. The adsorption of epoxy molecules in the bulk state from the ith layer to the $(i + 1)$ th, ith, and $(i - 1)$ th layers of the adsorbed state, which results in a decrease in $N_{E\infty,i}$.
4. The reacting of epoxy with amine in the ith layer of the bulk, which depletes $N_{E\infty,i}$.

The rate of change of epoxy $\frac{dN_{E\infty,i}}{dt}$ and amine $\frac{dN_{A\infty,i}}{dt}$ are thence represented by equations (4.11) and (4.15), respectively. Here, D_{EA} is the mutual diffusion coefficient in the binary epoxy–amine mixture, ΔL describes the physical thickness of a layer, and a, b, and c are represented by equations (4.12), (4.13), and (4.14), respectively, while d, e, and f are represented by equations (4.16), (4.17), and (4.18), respectively.

$$\frac{dN_{E\infty,i}}{dt} = D_{EA}\left(\frac{N_{E\infty,i-1} + N_{E\infty,i+1} - 2N_{E\infty,i}}{\Delta L^2}\right) + a - b - c \tag{4.11}$$

$$a = R_{d,E}(i, i - 1) + R_{d,E}(i, i) + R_{d,E}(i, i + 1) \tag{4.12}$$

$$b = R_{a,E}(i, i - 1) + R_{a,E}(i, i) + R_{a,E}(i, i + 1) \tag{4.13}$$

$$c = n_1 k_r N_{E\infty,i} \tag{4.14}$$

$$\frac{dN_{A\infty,i}}{dt} = D_{EA}\left(\frac{N_{A\infty,i-1} + N_{A\infty,i+1} - 2N_{A\infty,i}}{\Delta L^2}\right) + d - e - f \tag{4.15}$$

$$d = R_{d,A}(i, i - 1) + R_{d,A}(i, i) + R_{d,A}(i, i + 1) \tag{4.16}$$

$$e = R_{a,A}(i, i - 1) + R_{a,A}(i, i) + R_{a,A}(i, i + 1) \tag{4.17}$$

$$f = n_2 k_r N_{E\infty,i} \tag{4.18}$$

Finally, the rate of change of products due to reaction in the bulk state within the ith layer, $\frac{dN_{P\infty,i}}{dt}$, is expressed by equation (4.19).

$$\frac{dN_{P\infty,i}}{dt} = k_r N_{E\infty,i} \tag{4.19}$$

The mutual diffusion coefficient in the binary epoxy–amine mixture, D_{EA}, can be determined using equation (4.20), where D_0, f_g, and α_f are constants; E_D is the activation energy; b_D is an empirical constant; and $T_g(\xi)$ describes the free volume that is available and the degree of rotational restriction, and is defined by a ratio of the total epoxy concentration as a function of time, to the concentration at $t = 0$.

$$D_{EA} = D_0 \exp\left(-\frac{E_D}{RT}\right)\exp\left[-\frac{b_D}{(f_g + \alpha_f[T - T_g(\xi)])}\right] \tag{4.20}$$

The extent of cure, k_r, can be determined using equation (4.21) where k_{r0} is the Arrhenius pre-exponential constant, δ_0 is the coordination sphere reaction parameter, and E_a is the reaction rate activation energy.

$$k_r = \frac{k_{r0}\exp(-E_a/RT)}{1 + \dfrac{\delta_0}{D_{EA}}\exp(-E_a/RT)} \tag{4.21}$$

The rate equations for species in the adsorbed state are solved simultaneously as ordinary differential equations with those in the bulk state for each of the unknown parameters, $N_{E,i}$, $N_{A,i}$, $N_{P,i}$, $N_{E\infty,i}$, $N_{A\infty,i}$, and $N_{P\infty,i}$ in each of the layers.

4.2.2 Thermodynamics of formation

In Pitchumani's review [10] the thermodynamics of interphase formation is discussed in terms of reacting [13] fibre–matrix modelling, which was updated by Palmese [12] to cover the case of non-reacting species. The original model focusses on both the interface in terms of chain-to-surface modelling, and the interphase in terms of chain-to-chain modelling, once again using the epoxy–amine binary thermoset as an example.

Chain-to-surface interactions, $I_{cs,}$ are represented by equation (4.22), where $q(x_A, z)$ is calculated in accordance with equation (4.23) and x_A is the amine chain length, ω_A is the amine surface potential, k is the Boltzmann constant, T is temperature, z is the number of the lattice layer from the fibre surface, c is the characteristic constant bond length of C–C, and l is the characteristic constant bond length of the polymer chain. $S_A(z)$ in equation (4.22) is computed in accordance with equation (4.24).

$$I_{cs} = q(x_A, z)\exp[S_A(z)] \tag{4.22}$$

$$q(x_A, z) = 1 - \sqrt{\frac{2}{\pi x_A}}(1 - e^{\omega_A/kT})\text{erfc}\left(\frac{z}{\sqrt{c\frac{x_A l^2}{3}}}\right) \tag{4.23}$$

$$S_A(z) = \frac{\omega_A}{kT}\exp\left(\frac{\omega_A}{kT}\right)\sqrt{\frac{2}{\pi x_A}}\,\text{erfc}\left(\frac{z}{\sqrt{c\frac{x_A l^2}{3}}}\right) \tag{4.24}$$

Chain-to-chain interactions, I_{cc}, are represented by equation (4.25), where $v_E(z)$ is the volume fraction of epoxy, $v_E(\infty)$ is the volume fraction of epoxy in the far region, and χ is the interaction parameter for chain-to-chain interaction, such that two species repel one another when $\chi > 0$.

$$I_{cc} = \exp\left[\chi(v_E^2(\infty) - v_E^2(z))\right] \tag{4.25}$$

Excess mixing, M, is a phenomenon caused by different polymer chain lengths in the epoxy, x_E, and the amine, x_A. This is calculated in accordance with equation (4.26), where N_L is the number of lattice layers contained within a sublattice, where the thickness of each is l^0. Large values of N_L decrease the effect of excess mixing.

$$M = \exp\left[\frac{v_A(z) - v_A(\infty)}{N_L}\left(1 - \frac{x_A}{x_E}\right)\right] \tag{4.26}$$

Finally, the volume fraction of amine in the zth lattice layer, $v_A(z)$, is calculated from equation (4.27), where κ is a normalisation parameter to ensure that $v_A(z) + v_E(z) = 1$, and $v_A(\infty)$ is the volume fraction of amine in the far region.

$$v_A(z) = \kappa v_A(\infty) I_{cs} I_{cc} M \tag{4.27}$$

4.3 Mechanical properties and the interphase

4.3.1 Rule of Mixtures

The Rule of Mixtures is considered a good predictor of composite elastic modulus, E_c, properties in the axial direction of unidirectional continuous fibre-reinforced plastics, equation (4.28), for two phase composites. Here, E_f and E_m are the fibre and matrix elastic moduli, respectively, and V_f and V_m are the fibre and matrix volume fractions, respectively.

$$E_c = E_f V_f + E_m V_m \tag{4.28}$$

When the interphase plays a more critical role in the composite, this rule can be extended to a three-phase system, equation (4.29), where E_i and V_i are the elastic modulus and the volume fraction of the interphase, respectively. The volume fraction of interphase material can be as high as 15% at its upper limit [14], and in these cases there is a case that can be made for the incorporation of interphase contributions to the overall elastic modulus.

$$E_c = E_f V_f + E_m V_m + E_i V_i \tag{4.29}$$

Antifantis [14] suggested an empirical relationship for the determination of E_i, equation (4.30). Here, p is an adhesion factor calculated according to equation (4.31), where E_{io} represents the most extreme value of elastic modulus within the interphase, essentially the elastic modulus of the pinned polymer [15] region at the interface (at the fibre surface); ξ_1 represents the ratio of fibre and matrix elastic moduli, equation (4.32); r denotes the radial component of the cylindrical coordinate system; and r_f and r_i are the fibre and interphase radii, respectively.

$$E_i(r) = E_m\left[1 + (p\xi_1 - 1)\left(\frac{1 - re^{\frac{1-r/r_i}{r_i}}}{1 - r_f e^{\frac{1-r_f/r_i}{r_i}}}\right)\right] \tag{4.30}$$

$$p = \frac{E_{io}}{E_f} \tag{4.31}$$

$$\xi_1 = \frac{E_f}{E_m} \tag{4.32}$$

Antifantis [14] further suggested that the interphase Poisson's ratio could be predicted, ν_i, under similar principles using equation (4.33). Here, q is an adhesion factor calculated according to equation (4.34) and ξ_2 is expressed by equation (4.35) as a ratio of the fibre to matrix Poisson's ratios, ν_f and ν_m, respectively. ν_{io} represents the most extreme Poisson's ratio within the interphase, and is essentially correlated with the elastic modulus of the pinned polymer region at the interface (at the fibre surface).

$$\nu_i(r) = \nu_m\left[1 + (q\xi_2 - 1)\left(\frac{1 - re^{\frac{1-r/r_i}{r_i}}}{1 - r_f e^{\frac{1-r_f/r_i}{r_i}}}\right)\right] \tag{4.33}$$

$$q = \frac{\nu_{io}}{\nu_f} \tag{4.34}$$

$$\xi_2 = \frac{\nu_f}{\nu_m} \tag{4.35}$$

Equations (4.30) and (4.33) describe an interphase with a thickness t_i defined by the interphase radius less the fibre radius, equation (4.36). The elastic properties through this thickness vary exponentially as described by the exponential term in

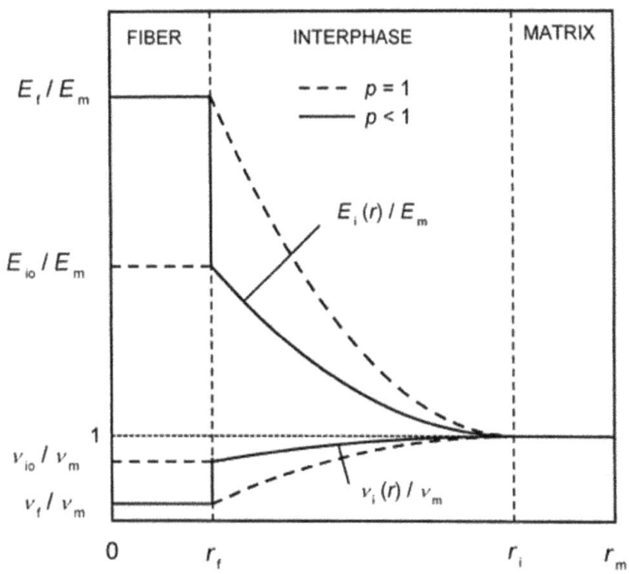

Figure 4.3. Variation of the elastic moduli and Poisson's ratios within the unit cell along the radius. Reprinted from [14]. Copyright (2000), with permission from Elsevier.

equations (4.30) and (4.33), i.e. $\left(\frac{1 - re^{\frac{1-r/r_i}{r_i}}}{1 - r_f e^{\frac{1-r_f/r_i}{r_i}}}\right)$ was chosen for the sake of generality rather than being borne from an empirical or semi-empirical basis. Figure 4.3 illustrates how the predicted elastic moduli and Poisson's ratios (normalised against the matrix) would thus vary from the fibre through the interphase and to the matrix.

$$t = \Delta r = r_i - r_f \tag{4.36}$$

4.3.2 Transverse Rule of Mixtures

The Transverse Rule of Mixtures as defined by the Reuss model is considered a good predictor of the transverse to the axial direction composite elastic modulus, $E_{c,t}$, of unidirectional continuous fibre-reinforced plastics, equation (4.37), for two-phase composites, since $\sigma_f = E\varepsilon_f$, $\sigma_m = E\varepsilon_m$. Here, σ_f is the fibre stress under elastic loading; ε_f and ε_m are the fibre and matrix strains under elastic loading, respectively; V_f and V_m are the volume fractions of the fibre and matrix, respectively; and E_f and E_m are the fibre and matrix elastic moduli, respectively.

$$E_{c,t} = \frac{\sigma_f}{V_f \varepsilon_f + V_m \varepsilon_m} = \left(\frac{V_f}{E_f} + \frac{V_m}{E_m}\right)^{-1} \tag{4.37}$$

When the interphase plays a more critical role in the composite, this rule could be applied, by extension, to a three-phase system, equation (4.38), where ε_i is the interphase strain under elastic loading. V_i is the interphase volume fraction, and E_i is the interphase elastic modulus. Here, $\sigma_f = E\varepsilon_f$, $\sigma_m = E\varepsilon_m$, $\sigma_i = E\varepsilon_i$. As previously mentioned, the volume fraction of interphase material can be as high as 15% at its upper limit [14], and in such instances, a case could be made for the incorporation of interphase contributions to the overall transverse composite elastic modulus.

$$E_{c,t} = \frac{\sigma_f}{V_f\varepsilon_f + V_m\varepsilon_m + V_i\varepsilon_i} = \left(\frac{V_f}{E_f} + \frac{V_m}{E_m} + \frac{V_i}{E_i}\right)^{-1} \tag{4.38}$$

The value for E_i can be estimated using one from a number of different models, including that shown in equation (4.30), or for example using a simpler power law expression suggested by Wacker [16, 17], equation (4.39), where $0 < \alpha < 1$ and $n = 2, 3, \ldots$.

$$E_i(r) = (\alpha E_f - E_m)\left[\frac{r_i - r}{r_i - r_f}\right]^n \tag{4.39}$$

The average transverse interphase modulus in itself, $E_{i,t}$, can separately be estimated using Wacker's model [16, 17], equation (4.40), where r denotes the radial component of the cylindrical coordinate system, and r_f and r_i are the fibre and interphase radii, respectively.

$$E_{i,t}(r) = \frac{r_i - r_f}{\displaystyle\int_{r_f}^{r_i} \frac{\mathrm{d}r}{E_i}(r)} \tag{4.40}$$

4.3.3 Modulus and strength models for nanocomposites

The extent to which interphase material will affect the mechanical properties of a composite is dependent on the fraction of interphase material relative to the fractions of the fibre and matrix materials. Since interphase thicknesses may reach up to 50 μm in thickness, nanoscale reinforcements are particularly affected by the presence of large volumes of interphase. This is because nanoscale reinforcements (defined as being < 100 nm in at least one direction) are, relatively speaking, small in size and as such, similarly sized interphase fractions might introduce a third reinforcing phase into the composite, with different reinforcing properties and behaviours. The design of nanocomposites should therefore where applicable take into account the interphase in design calculations. Mechanical design parameters such as the elastic modulus and composite strength have been detailed in [18], and each is covered below.

A model used to predict the elastic modulus in nanocomposites suggested by Ji and co-workers [19] has been experimentally validated and provides a good baseline for prediction as it covers a range of different reinforcement geometries. Reinforcement cross sections in this model were approximated to any of spherical, layered, or cylindrical geometries. The model, shown in equation (4.41), is shown for the ratio of composite elastic modulus, E_c, to that of the matrix material, E_m. In this equation, α and β are geometrical parameters associated with the interphase seen in equations (4.42), (4.43), and (4.44) and to the nanoparticle, equation (4.45), respectively. A ratio, m, of the interphase elastic modulus, E_i, and the matrix elastic modulus is shown in equation (4.46). E_p is the elastic modulus of the nanoparticle. α is calculated for spherical, layered, and cylindrical nanoparticles (see equations (4.42), (4.43), and (4.44)), where φ_f is the volume fraction of the nanoparticle component of the composites, r is the nanoparticle radius, d is the nanoparticle thickness or diameter, and t is the thickness of the interphase around the nanoparticle.

$$\frac{E_c}{E_m} = \left(1 - \alpha + \frac{\alpha - \beta}{1 - \alpha + \frac{\alpha(m-1)}{\ln m}} + \frac{\beta}{1 - \alpha + \frac{(\alpha - \beta)(m+1)}{2} + \beta\frac{E_{np}}{E_m}}\right)^{-1} \tag{4.41}$$

$$\alpha_{spherical} = \sqrt{\left(\frac{t}{r} + 1\right)^3 \varphi_f} \tag{4.42}$$

$$\alpha_{layered} = \sqrt{\left(2\frac{t}{d} + 1\right)\varphi_f} \tag{4.43}$$

$$\alpha_{\text{cylindrical}} = \sqrt{\left(\frac{t}{r} + 1\right)^2 \varphi_f} \tag{4.44}$$

$$\beta = \sqrt{\varphi_f} \tag{4.45}$$

$$m = \frac{E_i}{E_m} \tag{4.46}$$

φ_i represents the interphase volume fractions for different shaped nanoscale reinforcements, and this parameter can be calculated in accordance with equations, (4.47), (4.48), and (4.49), for spherical, layered, and cylindrical nanoscale reinforcements, respectively.

$$\varphi_{i,\text{spherical}} = \varphi_f\left[\left(\frac{r+t}{r}\right)^3 - 1\right] \tag{4.47}$$

$$\varphi_{i,\text{layered}} = \varphi_f\left(\frac{2t}{d}\right) \tag{4.48}$$

$$\varphi_{i,\text{cylindrical}} = \varphi_f\left[\left(\frac{r+t}{r}\right)^2 - 1\right] \tag{4.49}$$

$$\alpha = \sqrt{\varphi_i + \varphi_f} \tag{4.50}$$

The equations (4.47), (4.49), and (4.49) can be coupled to their equivalents for α shown in equations (4.42), (4.43), and (4.44), and it is clear that α can be represented by equation (4.50). This in turn can be entered into equation (4.41) for all types of nanocomposites [20].

Pukansky's model [21], equation (4.51), is a good baseline model for the prediction of nanocomposite tensile strength including the interphase. The closeness of fit for this model is validated against experimental data pertaining to a variety of differently shaped particles used as nanocomposite filler. The model was not intentionally designed for nanocomposites; however, when broken down, it is clear that the model construct is equally relevant at the nanoscale. The model parameters include σ_c and σ_m, which are the strengths of the composite and matrix, respectively; B, which is a parameter reflecting the interaction and adhesion of reinforcing and matrix; and φ_f, which is the reinforcement volume fraction.

$$\frac{\sigma_c}{\sigma_m} = \frac{1 - \varphi_f}{1 + 2.5\varphi_f} \exp(B\varphi_f) \tag{4.51}$$

$$B = (1 + A_{np}\rho_{np}t) \ln \frac{\sigma_i}{\sigma_m} \tag{4.52}$$

B is an empirical parameter defined by equation (4.52), where A_{np} is the nanoparticle reinforcement surface area, ρ_{np} is the nanoparticle reinforcement density, t is the interphase thickness, and σ_i is the interphase strength. Pukansky's model has application across a broad range of different types of nanocomposites [22].

References

[1] Drzal L T, Rich M J, Koenig M F and Lloyd P F 1983 Adhesion of graphite fibers to epoxy matrices: II the effect of fiber finish *J. Adhes.* **16** 133–52

[2] Jones F R 2010 A review of interphase formation and design in fibre-reinforced composites *J. Adhes. Sci. Technol.* **24** 171–202

[3] Qi Y, Jiang D, Ju S, Zhang J and Cui X 2019 Determining the interphase thickness and properties in carbon fiber reinforced fast and conventional curing epoxy matrix composites using peak force atomic force microscopy *Compos. Sci. Technol.* **184** 107877

[4] Downing T D, Kumar R, Cross W M, Kjerengtroen L and Kellar J J 2000 Determining the interphase thickness and properties in polymer matrix composites using phase imaging atomic force microscopy and nanoindentation *J. Adhes. Sci. Technol.* **14** 1801–12

[5] Munz M, Sturm H, Schulz E and Hinrichsen G 1998 The scanning force microscope as a tool for the detection of local mechanical properties within the interphase of fibre reinforced polymers *Composites A* **29** 1251–9

[6] Dominkovics Z, Hari J, Kovacs J, Fekete E and Pukanszky B 2011 Estimation of interphase thickness and properties in PP/layered silicate nanocomposites *Eur. Polym. J.* **47** 1765–74

[7] Mai K, Mader E and Muhle M 1998 Interphase characterization in composites with new non-destructive methods *Composites A* **29** 1111–9

[8] Riano L, Chailan J F and Joliff Y 2021 Evolution of effective mechanical and interphase properties during natural ageing of glass-fibre/epoxy composites using micromechanical approach *Compos. Struct.* **258** 113399

[9] Gu Y, Li M, Wang J and Zhang Z 2010 Characterization of the interphase in carbon fiber/polymer composites using a nanoscale dynamic mechanical imaging technique *Carbon* **48** 3229–35

[10] Pitchumani R 2012 Interphases in composites *Long Term Durability of Polymer Matrix Composites* ed K V Pochiraju (Berlin: Springer Science and Business Media LLC)

[11] Yang F and Pitchumani R 2002 A kinetics model for interphase formation in thermosetting matrix composites *J. Appl. Polym. Sci.* **89** 3220–36

[12] Palmese G R 1992 Origin and influence of interphase material property gradients in thermosetting composites *Report* CCM 92-25 Center for Composite Materials, University of Delaware, Newark, DE

[13] Hrivnak J 1997 Interphase formation in reacting systems Number 97-05, University of Delaware Center for Composite Materials, Newark, DE

[14] Antifandis N K 2000 Micromechanical stress analysis of closely packed fibrous composites *Compos. Sci. Technol.* **60** 1241–8

[15] Touaiti F, Pahlevan M, Nilsson R, Alam P, Toivakka M, Ansell M P and Wilen C E 2013 Impact of functionalised dispersing agents on the mechanical and viscoelastic properties of pigment coating *Prog. Org. Coat.* **76** 101–6

[16] Wacker G 1996 *Experimentell gestutzte Identifikation ausgewahlter Eigenschaften glasfaser-verstarkter Epoxidharze unter Bercksichtigung der Grenzschicht* Thesis at the IfW University of Kassel, Germany

[17] Wacker G, Bledzki A K and Chate A 1998 Effect of interphase on the transverse Young's modulus of glass/epoxy composites *Composites A* **29A** 619–26

[18] Alam P 2021 *Composites Engineering: An A–Z Guide* (Bristol: IOP Publishing)

[19] Ji X L, Jing J K, Jiang W and Jiang B Z 2004 Tensile modulus of polymer nanocomposites *Polym. Eng. Sci.* **42** 983–93

[20] Zare Y and Rhee K Y 2020 A simple technique for calculation of an interphase parameter and interphase modulus for multilayered interphase region in polymer nanocomposites via modeling of Young's modulus *Phys. Mesomech.* **23** 332–9

[21] Pukansky B 1990 Influence of interface interaction on the ultimate tensile properties of polymer composites *Composites* **21** 255–62

[22] Zare Y and Rhee K Y 2020 A simple and sensible equation for interphase potency in carbon nanotubes (CNT) reinforced nanocomposites *J. Mater. Res. Technol.* **9** 6488–96

Chapter 5

Surface treatment of reinforcing fibres

5.1 Introduction

The sizing of fibre involves the application of a homogenous thin coating to the fibre surface, either during the fibre manufacturing process, or, as part of a post-process surface treatment. Sizing chemistry and surface treatments are very much a 'black box' outside of industry. Yet, the application of sizing, coatings, or other surface treatments to reinforcing fibres prior to their application to composites is vitally important. Generally, sizing comprises water solute (which will be evaporated out prior to fibre use), film-forming agents, chemical coupling agents, and additives. The sizing is usually applied with the following beneficial aspects in mind:

1. Improvements in the intimacy of **fibre–matrix adhesion**
2. Improved **mechanical performance** of composites
3. Improvements in **processing and manufacture**:
    - (a) Provides lubrication for weaving fibres into fabrics
    - (b) Decreased fibre fragmentation during processing
    - (c) Provides lubrication for ease of unwinding
    - (d) Decreased fibre agglomeration during manufacture and consolidation
    - (e) Reduced wear of blades during fibre chopping operations
4. Development of optimal **interphase** structures and conditions
5. **Chemical protection** including:
    - (a) Corrosion resistance
    - (b) Oil resistance
6. **Physical protection** during:
    - (a) Handling
    - (b) Moulding
    - (c) Compunding
    - (d) Processing

doi:10.1088/978-0-7503-5688-6ch5

7. **Environmental protection** including:
   (a) Thermal
   (b) Hydrolytic

A good sizing should typically seek to maximise fibre wetting, as this improves interfacial adhesion and encourages regularity in the formation of interphases. Good quality wetting will usually create the conditions for the partial solubility of sizing in a reactive solvent, and as such, it can be useful to determine the solubility interaction parameter, χ, equation (5.1), where V is molar volume, R is the universal gas constant, T is temperature, δ_P is the solubility parameter of polymer, and δ_S is the solubility parameter of the reactive solvent [2], each of which can be calculated using the group contribution method [3, 4]. The interaction parameter, χ, is deemed to show good interaction miscibility of the polymer in the solvent when $\chi < 0.5$ [2].

$$\chi = \frac{V}{RT}(\delta_P - \delta_S)^2 \tag{5.1}$$

Two of the more common mechanical test methods that are used when determining the effects of sizing on composite properties include interlaminar shear strength (ILSS) testing and interfacial shear strength (IFSS) testing. More detailed discourse on these and other mechanical test methods is provided in chapter 7. The ILSS method (covered in more detail in chapter 9 section 9.6) is usually based on standardised testing methods such as those in ASTM D2344/D2344M-22 [1], which is based on the 3-point bend testing of short beams (i.e. beams with an aspect ratio of 5 or less). The ILSS can be calculated using equation (5.2), where P is the maximum load imposed on the beam and b and h are the width and the thickness of beam, respectively.

$$\text{ILSS} = \frac{3P}{4bh} \tag{5.2}$$

The IFSS is measured using the microbond test (see chapter 7 section 7.3.2). Here, a fibre is embedded into a resin droplet and subsequently pulled out. The force measured to pull the fibre out from the droplet is then applied to equation (5.3) to provide the shear strength value at the fibre–matrix interface. In this equation, F is the pullout force, d is the fibre diameter, and l is the embedding length.

$$\text{IFSS} = \frac{F}{\pi dl} \tag{5.3}$$

The two methods are contrasting since the IFSS measurement is significantly more local to the fibre–matrix interface, while the ILSS imparts load, which is distributed within the full composite structure and the shear stress failure is intended at a laminate–laminate interface. In addition, the ILSS does not always exhibit laminate–laminate failure behaviour, but may also show signs of plastic deformation and flexure-based failure (see chapter 9 section 9.6).

5.2 Carbon fibres

5.2.1 Common carbon fibre sizing

Carbon fibre sizing chemistry is very broad and includes compounds based on epoxies, polyesters, polyethers, acrylics, polyurethanes, and styrenes. Examples may include bisphenol A glycidyl ether (DGEBA) epoxy [5], figure 5.1, which can be fluorinated (covalently grafted with fluorine atoms) to enable hydrophobic behaviour and increase surface roughness and wettability [18]. This potentialises additional mechanical benefits in terms of the area available for adhesion and improvements in mechanical interlocking. DGEBA also improves the end properties of the sizing after treatment with methyl tetrahydrophthalic anhydride (MeTHPA) as it has been shown to increase the IFSS, the toughness, and the ILSS [19], as it enables excellent wettability at the fibre surface [20]. Glycidyl amine epoxies such as tetraglycidylether of 4, 4′-diaminodiphenyl methane (TGDDM) [6], figure 5.2, are often reacted with other compounds, examples of which may include 4, 4′-diaminodiphenylsulphone (DDS) [7] or dodecylamine ($H_2N{-}C_{12}H_{25}$) [8]. Other examples of common sizing include unsaturated polyester, figure 5.3, and polyether copolymer [9], figure 5.4. Maleic anhydride (MA) is a typical component in unsaturated polyester compounds and there is good evidence of its contribution to improved interfacial adhesion between fibre and matrix in polypropylene matrix composites [11–13], and in carbon/polyester composites [13]. If combined with o-phthalic anhydride, the polymer can be made stiffer and exhibits improved compatibility with styrene-based polymers [13]. Adjustments to sizing chemistry and matrix formulations are often needed to improve carbon fibre compatibility with the matrix polymer. Polypropylene, a non-polar polymer, is often functionalised by MA-based coupling agents, and when incorporating carbon fibres, unsizing the fibres can in fact improve interfacial adhesion [15]. This, in turn, is presumably attributable to the removal of polar sizing to reveal non-polar carbon fibre surfaces.

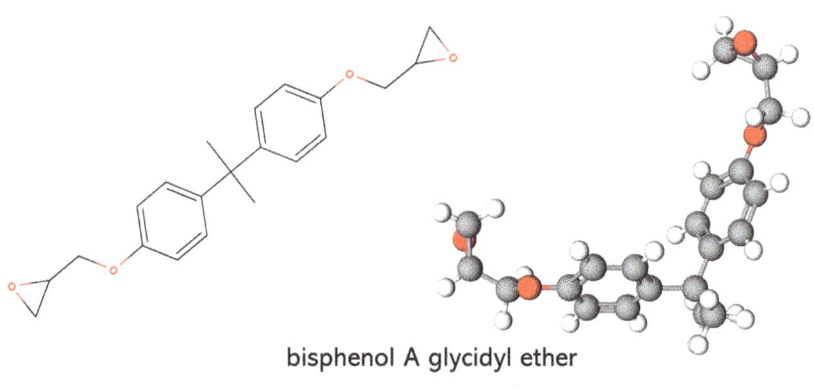

Figure 5.1. 2D and 3D chemical renders of an example bisphenol A glycidyl ether, a carbon fibre sizing.

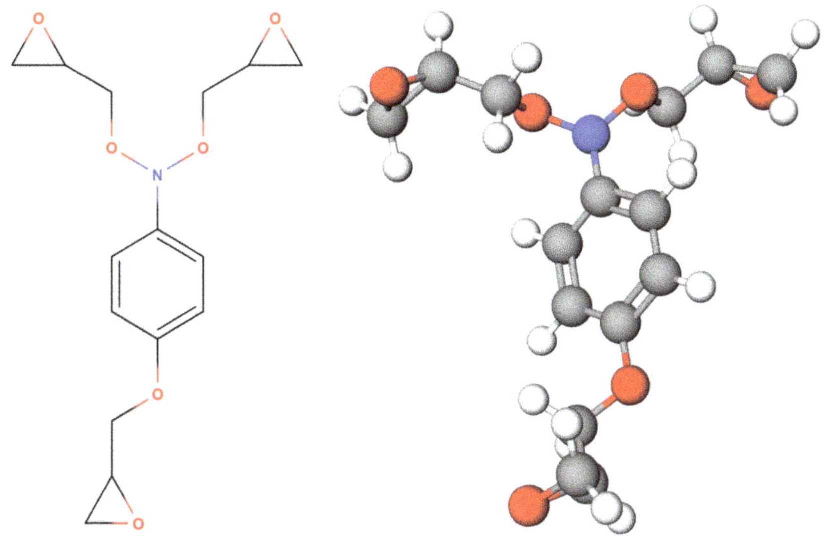

tetraglycidylether of 4,4′-diaminodiphenyl methane (TGDDM)

Figure 5.2. 2D and 3D chemical renders of an example of a glycidyl amine epoxy carbon fibre sizing, in this case tetraglycidylether of 4, 4′-diaminodiphenyl methane (TGDDM).

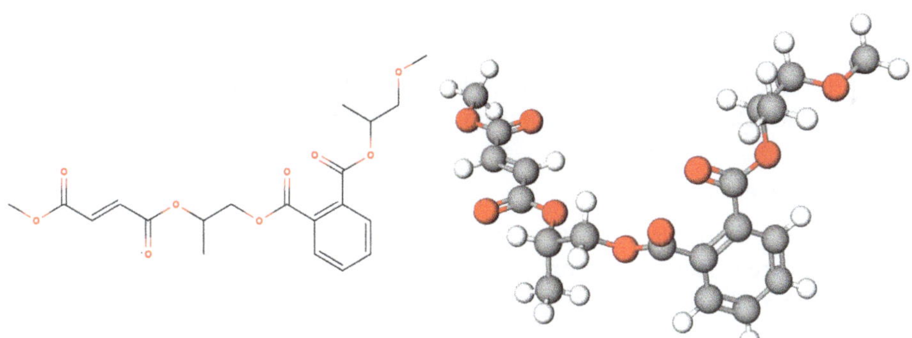

unstaurated polyester

Figure 5.3. 2D and 3D chemical renders of an example unsaturated polyester, a carbon fibre sizing.

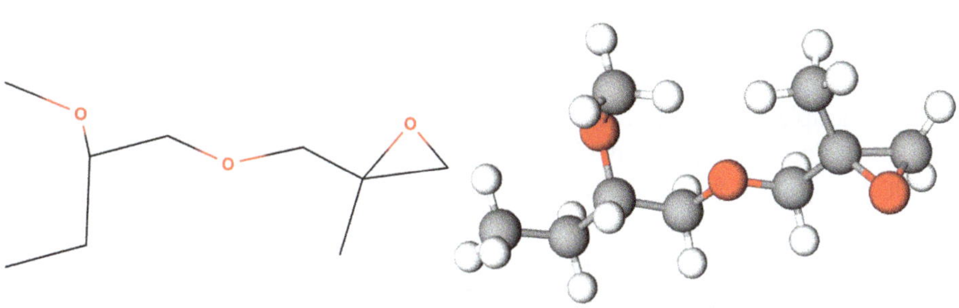

polyether copolymer (recurring unit)

Figure 5.4. 2D and 3D chemical renders of an example polyether copolymer, a carbon fibre sizing.

5.2.2 Sizing identification by IR spectroscopy

IR spectroscopy methods (discussed more in chapter 2) can be used to differentiate between sizing type and between sized and unsized carbon fibres. DEGBA exhibits bands in the mid-IR range including [O—H] stretching at ca. 3400 cm^{-1}; [C—H] stretching of the oxirane ring at 3039 cm^{-1} as well as [C—O] stretching of this ring at 933 cm^{-1} and 825 cm^{-1}; aromatic (CH_2) and aliphatic (CH) [C—H] stretching between 2965 and 3871 cm^{-1}; aromatic ring stretching of [C=C] at 1581 cm^{-1} and 1606 cm^{-1} as well as [C—C] stretching of the aromatic ring at 1506 cm^{-1}, [C—O—C] aryl alkyl ether stretches at 1036 cm^{-1}, 1180 cm^{-1}; 1230 cm^{-1}, and 1294 cm^{-1}; and [C—OH] bending of alcohols at 1011 cm^{-1}, 1086 cm^{-1}, and 1105 cm^{-1} [18]. With glycidyl amine epoxies such as tetraglycidylether of 4, 4′-diaminodiphenyl methane (TGDDM)/4, 4′-diaminodiphenylsulphone (DDS) compounds, the main near-IR (NIR) spectral bands can be expected at 9763 cm^{-1} and 6684 cm^{-1} ([NH_2] and [NH]) with [NH_2] also at 6578 cm^{-1} and 5067 cm^{-1}, 8627 cm^{-1} [CH—O—CH_2], [—OH] at both 6970 cm^{-1} and at 4897 cm^{-1}, and [—CH] and [CH_2] at 5882 cm^{-1} and 5232 cm^{-1} [7]. The chemical modification of sizing such as TGDDM most typically shows minor variations in spectral detail. Unmodified TGDDM by FTIR spectroscopy shows a characteristic hydroxyl absorption band at 3500 cm^{-1} and epoxide bands at 906 cm^{-1} and 754 cm^{-1}. When TGDDM is modified by DDS and dodecylamine [8], the hydroxyl absorption band shifts to 3450 cm^{-1} while the epoxide bands at 906 cm^{-1} and 754 cm^{-1} are still present, albeit with a reduced level of intensity. Important FTIR peaks that provide evidence for chemical bonding between unsaturated polyester sizing and carbon fibre are found at 1370 cm^{-1} and 1723 cm^{-1}, which indicate the formation of [C=O] bonding [14].

5.2.3 Mechanical properties

There is strong evidence to show that sizing improves the interfacial strength of adhesion of fibre to matrix. For example, Geng *et al.* [14] report improvements in the ILSS and IFSS of unsaturated polyester-sized carbon fibres in a vinyl ester matrix at levels of 18.8% and 43.4%, respectively. In their work they focussed on developing self-emulsifying anionic polyester sizing agents to promote adhesion between carbon and vinyl ester. The main reason for the increases in ILSS and IFSS is due to the unsaturated bonds in the polyester sizing. These bonds essentially react with free radicals on the carbon fibre surface and to enable chemical bonding between the two. However, it should be noted that a higher thickness sizing does not necessarily always improve the mechanical performance of the composite. Vauitard and Drzal [22], for example, report that the ILSS can rise initially and subsequently plateau beyond a certain sizing thickness, and that other properties such as 90 ° flexural strength may in fact reach an apex at a certain coating thickness, after which the properties are reduced. While sizing overall has a positive effect on interfacial strength, there are some recorded cases where a decrease is reported, a phenomenon that can be explained by sizing effects on the fibre surface consistency, which in turn can be correlated with wettability (fibres with low adhesion exhibiting lower characteristics of wettability) [21]. This said, unsizing carbon fibres exposes their non-polar surfaces, which can be of benefit when strong adhesion is required at interfaces with non-polar matrix polymers such as polypropylene (PP). Improvements in mechanical performance through unsizing such fibres are reported at >10%, which can increase to up to 19% with the addition of MA [15]. Harper *et al.* [16] report improvements of 33% and 34% in the tensile and flexural moduli, respectively, of MA/PP-sized carbon fibre-reinforced PP over PP-sized carbon fibre-reinforced PP. The importance of sizing fibres with polar or non-polar compounds based on the matrix material can be demonstrated by comparing the effect of different fibre surface polarities against polar polycarbonate (PC) matrix polymers. Unsized carbon fibre is non-polar and as such has a lower binding affinity to PC than carbon fibres sized with epoxy/phenoxy (EPOPHE) and phenoxy (PHE), affecting its comparative strength and modulus as demonstrated by Ozkan *et al.* [17].

5.3 Glass fibres

Glass fibres used to reinforce plastics are essentially silica-based materials, though there are differing compositions of fibres to suit a broad range of applications. The most common of these is E-glass, or electrical glass (with high electrical resistivity), which exhibits excellent Young's modulus, $E = 72.3$ GPa, and tensile strength, $\sigma_t = 3.44$ GPa, values. These fibres are widely used in aerospace, automotive, renewable energy technology, and marine industries. They are essentially alkali glasses containing silica, alumina, and trace oxides of metals such as calcium, magnesium, boron, sodium, and potassium. Another high-performance reinforcing glass fibre is S-glass. This is a high-strength, high-cost glass used in aerospace industries and in military ballistic technologies. It has a Young's modulus of $E = 85.5$ GPa and a tensile strength of $\sigma_t = 3.45$ GPa [23]. Aside from these, there is A-glass (alkali glass), AR-glass (alkali resistant glass), C-glass (chemical glass), D-glass (low dielectric glass), ECR-glass (electronic glass), and S_2-glass (silica glass), more detail for each of which can be found in [23].

5.3.1 Common glass fibre sizing

Glass fibres are typically sized with organofunctional silanes, which have the generic formula $X–Si(OR)_3$, where R is either an ethyl or methyl group and X is any of the following functional groups: amino, epoxy, methacryloxy, or vinyl. According to Thomason's book on size chemistry [24], a vast majority of glass sizing types can be narrowed down to those that contain γ-(aminopropyl)triethoxysilane or 'APTES' (molecular weight: 220.36 Da, formula: $C_9H_{22}NO_3Si$); those containing γ-(glycidoxypropyl)trimethoxysilane or 'GPTMS' (molecular weight: 236.34 Da, formula: $C_9H_{20}O_5Si$); those containing γ-methacryloxypropyltrimethoxysilane or 'MPTMS' (molecular weight: 248.35 Da, formula: $C_{10}H_{20}O_5Si$); or those that contain vinyltriethoxysilane or 'VTES' (molecular weight: 190.31 Da, formula: $C_8H_{18}O_3Si$). Both 2D and 3D chemical renders of APTES, GPTMS, MPTMS, and VTES are illustrated in figure 5.5. Of these, APTES is the most commonly patented, researched, and used [29].

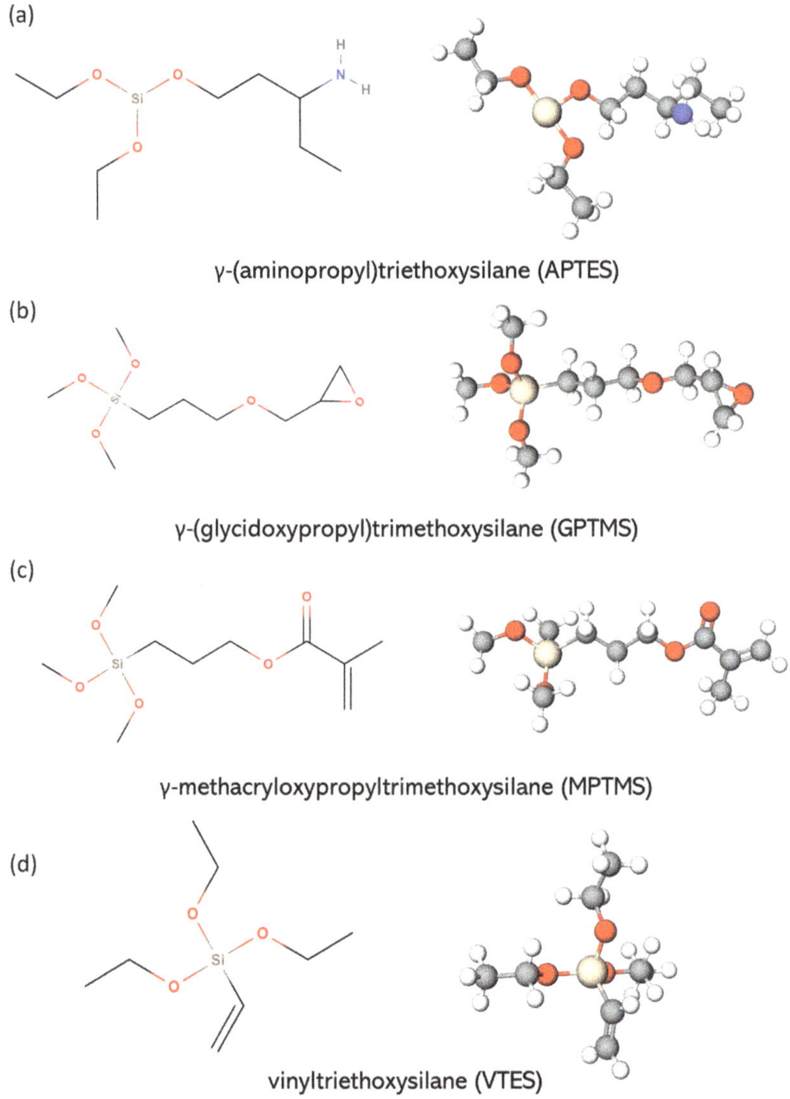

Figure 5.5. 2D and 3D chemical renders of oft-used glass fibre sizing showing (a) γ-aminopropyltriethoxysilane (APTES), (b) γ-glycidoxypropyltrimethoxysilane (GPTMS), (c) γ-methacryloxypropyltrimethoxysilane (MPTMS), and (d) vinyltriethoxysilane (VTES).

5.3.2 Sizing identification by IR spectroscopy

IR spectroscopy methods (discussed more in terms of FTIR in chapter 2) can be used to differentiate between sizing type and between sized and unsized glass fibres. Figure 5.6 from [30] shows FTIR spectra of APTES, MPTMS, and unsized silica glass. Here, both APTES and MPTMS show distinct peaks at 3374 cm^{-1} and at 2564 cm^{-1}, indicating the presence of N—H bonding and S—H bonding, respectively. Peaks at 2973 cm^{-1} and 2883 cm^{-1} indicate the CH_2 and CH_3 moities from

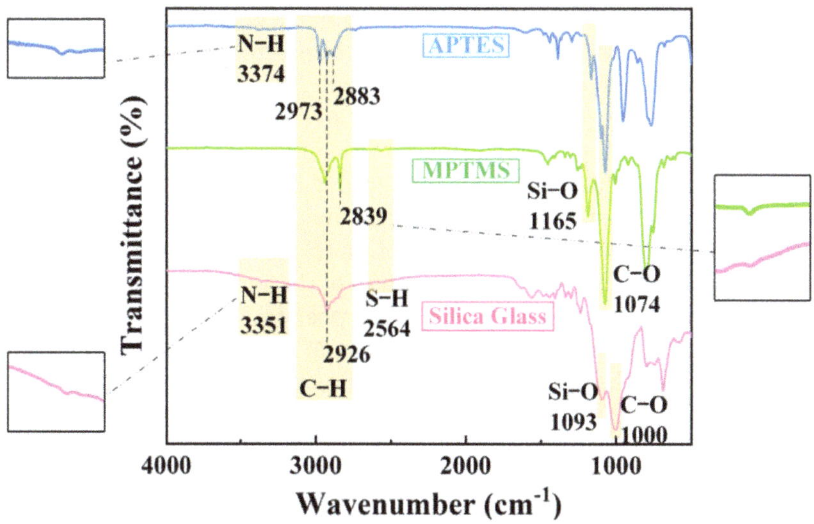

Figure 5.6. FTIR spectra of APTES, MPTMS, and silica glass. Reproduced from [30].

Si—OCH_2CH_3 in APTES, respectively. The CH_3 moiety from -$(CH_2)_3SH$ in MPTMS can be noted from the absorption peak at 2839 cm^{-1}. The absence of the CH_3 peak in silica is evidence of hydrolytic condensation of ethyoxyl and methoxyl resulting from APTES and MPTMS reactions, which create dense Si—O—Si networks through high levels of cross-linking. Finally, the absorption peaks at 2926 cm^{-1} after reaction with silica come from the -$(CH_2)_3NH_2$ in APTES and -$(CH_2)_3SH$ in MPTMS.

Untreated E-glass displays an absorption peak at 900 cm^{-1}, indicative of [Si—O—Si] bond [34]. Aminosilanes including 3-aminopropyltrimethoxysilane (APTMS) [32] and (3-aminopropyl) triethoxysilane (APTES) [34], which are used as silane coupling agents in the process of silanisation, show [—CH] stretching of the propyl groups at 2900 cm^{-1}, 2915 $cm^{-1,}$ and 3003 cm^{-1}; [—CH] bending of the propyl groups at 1490 cm^{-1}; and [NH_2] stretching from the presence of amine groups in aminosilanes at 3432 cm^{-1} [32, 34].

5.3.3 Mechanical properties

This is strong evidence for mechanical benefit in sizing glass fibres. While the use of aminosilanes is very common [31], additional coupling agents such as low-molecular-weight polypropylenes have been shown to improve the flexural properties by changing the adhesion and morphology of the interphase [33]. The use of aminosilanes has nevertheless been shown to provide significant mechanical improvement over acid and base-treated glass fibres. Comparing the effect of glass fibre surface treatments, Arunprakash and Rajadurai [32] report that composites containing fibres sized using 3-aminopropyletrimethoxysilane (APTMS) and silanised yielded ILSS improvements of up to 12.5%, while the ILSS of H_2SO_4 (acid)-treated and NaOH (base)-treated fibre composites were at the lowest, 24% reduced from control

(APTMS sized) samples. Kiss and co-workers [35] report a 1.88-fold improvement in the compressive strength of aminosilane (aminopropyltriethoxysilane)-sized glass fibre yarn-reinforced PA6 over unsized equivalents, which after a 50J impaction was still 1.85-fold higher strength in compression than post-impacted unsized equivalent composites. There are three consequences of sizing glass fibre that result in higher IFSS values. These include the effect the sizing has on the interphase [36] with mechanically superior sizing typically cross-linking with the matrix to form interphase matter, the energy absorption prior to initial debonding, and the post-debond sliding energy, which is itself a function of the surface roughness of the sizing on the fibre [37]. Hygrothermal ageing is a phenomenon now known to affect composites by altering the properties of matrix matter (swelling, plasticisation, etc.), and by also attacking the interfaces between fibre and matrix [38]. These interfacial effects can be evidenced by comparing the IFSS of hygrothermally aged interfaces against those that are dry unaged. For example, Benethuiliere and co-workers [2] compared the IFSS of glass fibre-reinforced vinyl ester composites with 1.86% by weight polyvinylacetate sizing, subjected to ageing at 70 °/95%RH/7-days, 70 °/95% RH/14-days, and as dry material. In their work they report a 22% drop in IFSS in styrene-based vinyl ester matrix composites for the longest periods of ageing. However, they also note that replacing styrene with butanediol dimethacrylate (BDDMA) in vinyl ester matrix composites liberates interfaces from hygrothermally induced damage, reporting no statistical differences in the IFSS of BDDMA-containing composites under both aged and dry-unaged conditions.

5.4 Natural fibres

There has been an increased impetus in the research and development of natural fibre-reinforced polymer (NFRP) composites over recent decades as they offer a renewable, sustainable, and biodegradable alternative to common engineering fibres such as glass, carbon, aramid, boron, and others. There are numerous challenges in using natural fibre reinforcements including issues related to hygroscopic properties, incompatibility with matrix matter, fire damage tolerance, and mechanical performance. As such, there is a broad range of biological (bacterial, fungal, enzumatic), chemical (alkalisation, acetylation, coupling agents, oxidisation), and physical treatments (plasma, steam explosion, thermo-mechanical) [39] that can be applied to improve the fibre surface, many of which specifically target improvements to the compatibility between fibre and matrix material. This section will briefly focus on describing a few common treatment types applicable to natural fibre reinforcements, and will highlight the effects of fibre treatment on the physical and/or mechanical properties of NFRP composites.

5.4.1 Alkalisation

Alkalisation (also referred to as mercerisation) uses a strong base, often sodium hydroxide (NaOH), to partially or completely clean impurities, lignins, and hemicelluloses from the surface of a natural plant fibre. The immersion of natural fibres into a base has additional impacts to the structure, dimensions, and morphology of the fibre [40]. The base concentration impacts the extent to which lignins and hemicelluloses are removed [41], and when these are removed from the fibre, the microfibril angle aligns towards the fibre axis [42, 43] while there is a concurrent positive disruption of intra-fibrillar hydrogen bonding [41, 44], resulting in improved fibre tensile strength and stiffness. The effects of microfibril angle can be determined through single fibre tensile testing using equation (5.4), where ε is the strain at break, ΔL is the elongation at break, L_0 is the fibre gauge length, and α is the microfibril angle (in degrees) [43].

$$\varepsilon = \ln\left(1 + \frac{\Delta L}{L_0}\right) = -\ln(\cos\alpha) \tag{5.4}$$

Alkalisation at high base concentrations can actually be detrimental to the properties of the alkalised fibre [45–48], and as such, it can be useful to have a 'safe' concentration range within which one can be sure to improve fibre properties and performance. High base concentrations essentially depolymerise the cellulose polysaccharide, whilst excessively delignifying the fibre, leading to weakened non-continuum fibrillar structures. Many outputs indicate that a reliable NaOH concentration for the alkalisation of plant fibre is between 0.03 and 10%. There are several works highlighting the effect of alkalisation within this range on both the mechanical properties of individual natural plant fibres and the final properties of NFRP composites. A few examples are summarised in tables 5.1 and 5.2, respectively.

Natural fibres often show high variability in surface roughness as a result of alkalisation, and as such, are very sensitive to the size of the spot area over which physical measurements (e.g. roughness) are taken. For example, George and co-workers [58] provide evidence of proportional increases in *RMS* roughness as a function of area spot size on untreated and treated hemp fibres, as shown in figures 5.7(a)–(c). Figure 5.7 shows variations in *RMS* roughness plotted against spot area size for (a) hemp fibres treated with NaOH (sodium hydroxide) and

Table 5.1. Examples: natural fibres subjected to alkalisation and the effects on fibre mechanical properties. Here, σ_t is tensile strength and E_t is the tensile modulus. Only optimal improvements shown.

Description	Source
Sansevieria trifasciata fibre (σ_t = 483 MPa, E_t = 3.70 GPa) after 5% NaOH alkalisation (σ_t = 535 MPa, E_t = 4.23 GPa)	[41]
Hemp fibre ($\sigma_t \approx$ 275 MPa, $E_t \approx$ 9.5 GPa) after 5% NaOH alkalisation ($\sigma_t \approx$ 300 MPa, $E_t \approx$ 15 GPa)	[39]
Hemp fibre bundles (σ_t = 594 MPa, E_t = 37.5 GPa) after 0.24% NaOH alkalisation (σ_t = 1074MPa, E_t = 58.1 GPa)	[49]
Sisal fibre bundles (σ_t = 500 MPa, E_t = 26.4 GPa) after 0.16% NaOH alkalisation (σ_t = 820 MPa, E_t = 41.4 GPa)	[50]
Symphirema involucratum stem fibre (σ_t = 397.2 MPa, E_t = 4.6 GPa) after 5% NaOH alkalisation (σ_t = 471.2 MPa, E_t = 5.8 GPa)	[51]

Table 5.2. Examples: natural fibres subjected to alkalisation and the effects on NFRP composite mechanical properties. Here, σ_t is tensile strength and E_t is the tensile modulus, σ_f is flexural strength, and E_f is the flexural modulus. Only optimal improvements shown.

Description	Source
New cane fibre (*Arundo donax L.*)/polyester matrix (σ_t = 16.3 MPa, E_t = 2.4 GPa) after 6% NaOH alkalisation (σ_t = 25.6 MPa, E_t = 2.9 GPa)	[53]
Jute fibre/epoxy matrix (σ_t = 39.8 MPa, E_t = 3.4 GPa) after 2% NaOH alkalisation (σ_t = 39.1 MPa, E_t = 3.6 GPa)	[54]
Jute/sisal fibre/epoxy matrix (σ_t = 66.8 MPa, E_t = 4.4 GPa) after 2% NaOH alkalisation (σ_t = 74.8 MPa, E_t = 6.76 GPa)	[54]
Ramie fibre/polylactic acid (PLA) matrix (σ_t = 39.3 MPa, E_t = 168.2 MPa) after 8% NaOH alkalisation (σ_t = 57.4 MPa, E_t = 248.3 MPa)	[55]
Bamboo fibre/polyester matrix ($\sigma_t \approx$ 19 MPa) after 6% NaOH alkalisation ($\sigma_t \approx$ 21 MPa)	[56]
Flax fibre/epoxy matrix (σ_f = 218 MPa, E_f = 18 GPa) after 3% NaOH alkalisation (σ_f = 283 MPa, E_f = 22 GPa)	[57]
Alfa fibre (*Stippa tenacissima*)/polyester matrix (σ_f = 22 MPa, E_f = 1.2 GPa) after 10% NaOH alkalisation (σ_f = 58 MPa, E_f = 3 GPa)	[52]

xylanase (Xyl); (b) NaOH (sodium hydroxide) and laccase (Lac); and (c) hemp fibres treated with derivatives of sulfonic acids: aniline-2-sulfonic acid (A2S) and 4-aminotoulene-3-sulfonic acid (AT3S).

5.4.2 Acetylation

Acetylation is a chemical modification process where an acetyl group [$CH_3C{=}O$] replaces a hydrogen atom to form acetate esters, or acetates. The chemical modification is hence an esterification reaction with acetic acid, and there are different chemical reactants that can be used to enable acetyl substitution with hydrogen. This is illustrated in figure 5.8, where acetic anhydride in reaction with OH attached to a plant cell wall surface causes the substitution of H with an acetyl group and residual acetic acid. While acetic anhydride is a common reactant, propionic anhydride can also be used to acetylate a fibre surface [59], as can vinyl acetate [60]. The acetylation process typically requires a catalyst such as potassium carbonate or pyridine.

One of the major benefits of acetylation is that it reduces susceptibility to moisture absorption by creating a hydrophobic surface, which concurrently improves its dimensional stability, removes the waxy layer from the fibre surface, and increases its tolerance to environmental degradation [61]. The formation of strong covalent bonds between the acetyl groups and the cell wall surface is understood to contribute to notable increases in strength and stiffness of acetylated fibre-reinforced composites [62]. Khalil and co-workers [63] compared the IFSS of untreated and acetylated oil palm empty fruit bunch (EFB) fibres as well as coir fibres in a range of different matrices. They report that actylated fibres showed overall improved IFSS over

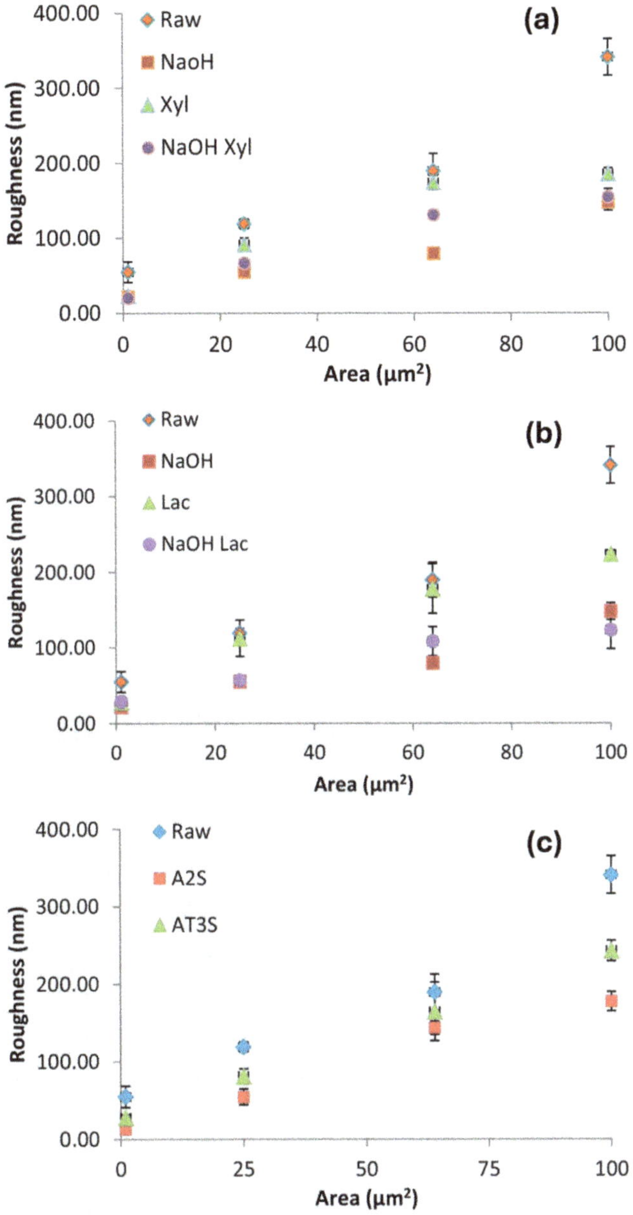

Figure 5.7. Surface roughness (*RMS*) plotted against spot area for hemp fibres treated with (a) NaOH (sodium hydroxide) and xylanase (Xyl), (b) NaOH (sodium hydroxide) and laccase (Lac), and (c) derivatives of sulfonic acids (A2S = aniline-2-sulfonic acid and AT3S = 4-aminotoulene-3-sulfonic acid). Reprinted from [58]. Copyright (2014), with permission from Elsevier.

untreated fibres in the majority of matrix materials tested, including Epiglass (commercial epoxy), West system (commercial epoxy), Crystic (unsaturated polyester), Metset (unsaturated polyester), and polystyrene, figures 5.9(a) and (b), with

Figure 5.8. Reaction of acetic anhydride at a cell wall (thick black line) showing the substitution of hydrogen from an OH with an acetyl group and residual acetic acid.

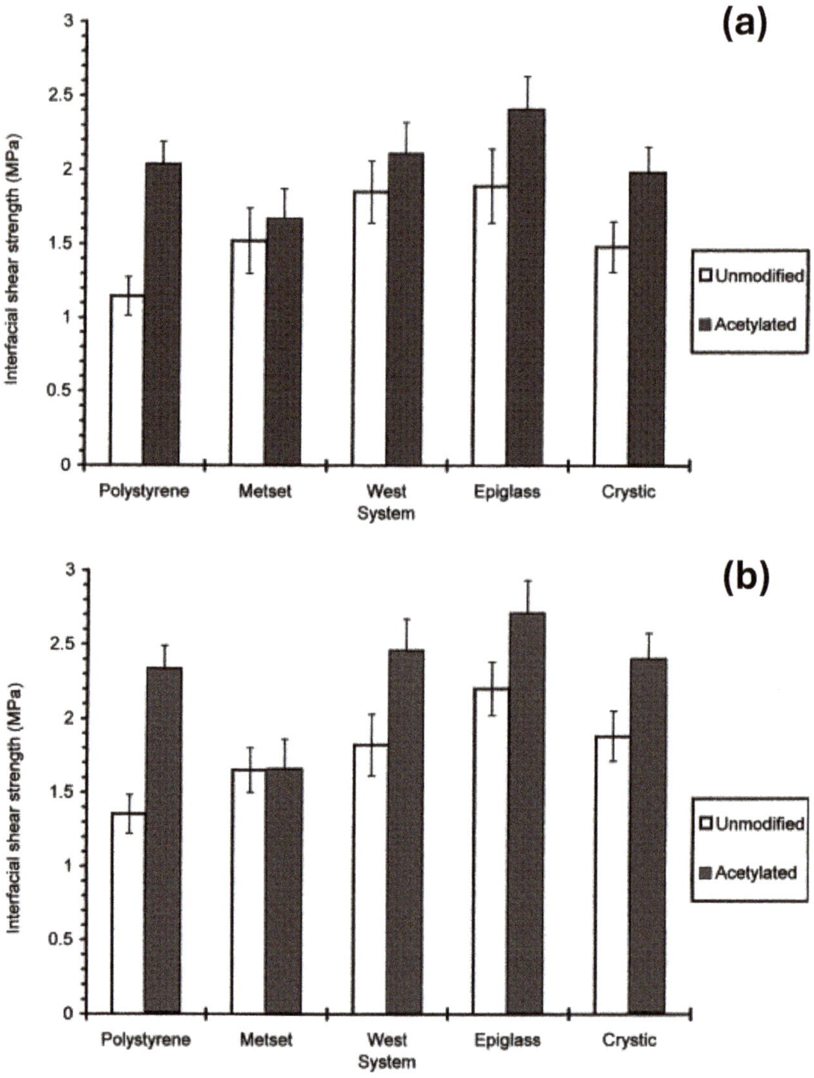

Figure 5.9. (a) IFSS of EFB and (b) IFSS of coir, each in the following matrix materials: Epiglass (commercial epoxy), West system (commercial epoxy), Crystic (unsaturated polyester), Metset (unsaturated polyester), and polystyrene. Error bars indicate standard deviations. Reprinted from [63]. Copyright (2001), with permission from Elsevier.

the superior improvements being noted to occur in hydrophobic resins, as acetylation increases the hydrophobicity of the fibre surfaces.

5.4.3 Silane coupling agents

Polymers and inorganic compounds can be attached to natural plant fibre surfaces to modify the surface properties of the fibres. There are numerous types of coupling agents that can be effectively grafted to the surfaces of fibres, and these are typically split into different classes depending on their chemical compositions and structures.

Silanes are an example of a specific class of silicon-containing coupling agent, the general chemical structure of which is $R_{4-n} - Si - (R'X)_n$, where R is alkoxy, X defines an organofunctional section, and R' is an alkyl bridge that connects the silicon atom with the organofunctional part [64]. The process of attaching silane coupling agents to natural fibre surfaces is termed silanisation, and the primary purpose is to more efficiently compatibilise natural fibre surfaces with matrix matter. A consequence of accomplishing this is an increase in the fibre–matrix strength of adhesion. A range of silane coupling agents are shown in table 5.3. This table provides detail on the different types of silane typically used in functionalising NFRP composites and details their chemical structures, organofunctionalities, and their targeted polymer matrices.

Silane coupling agents interact with natural fibre surfaces through various steps including hydrolysis, self-condensation, adsorption, and chemical grafting. These step-by-step interactions are chemically illustrated by Xie and co-workers [64] and shown in figure 5.10. During hydrolysis, figure 5.10(a), silane monomers are hydrolysed in water with either an acid or base catalyst. The reaction results in the production of silanol $R' - Si(OH)_3$. The next interaction step involves self-condensation, figure 5.10(b), where the silanols condense. Silanols should ideally be left free (uncondensed) during this step so they may adsorb to the hydroxyl groups on the surface of the natural fibres and a low pH environment is known to both accelerate the rate of hydrolysis, whilst concurrently reducing the extent of silanol condensation. In the third step, uncondensed silanol will adsorb to the hydroxyls present on the surfaces of the natural fibres through hydrogen bonding, figure 5.10 (c). Free silanols then adsorb and react with one another to form stable [Si—O—Si] bonds linking together polysiloxanes. Finally, hydrogen bonds between the silanol groups and the surface hydroxyls are converted to covalent [Si—O—C] bonds through heat exposure, figure 5.10(d).

5.4.4 Maleic anhydride coupling agents

Another common coupling agent used to functionalise natural fibres is MA. MA is typically grafted with a polymer to compatibilise it with the matrix material. Common examples include MA-grafted polypropylene (MA-g-PP/PP-g-MA) and MA-grafted polylactic acid (MA-g-PLA/PLA-g-MA). Similarly to silane coupling agents, MA coupling agents form strong covalent bonds with the fibre surface by substitution reaction with free hydroxyl groups, figure 5.11.

Table 5.3. Silanes used for the natural fibre/polymer composites: chemical structures, organofunctionalities, and target polymer matrices. Reprinted from [64]. Copyright (2010), with permission from Elsevier.

Structure	Functionality	Abbreviation	Target matrix
$(RO)_3Si{-}(CH_2)_3{-}NH_2$[a]	Amino	APS	Epoxy Polyethylene Butyl rubber Polyacrylate PVC
$(RO)_3Si{-}CH{=}CH_2$	Vinyl	VTS	Polyethylene Polypropylene Polyacrylate
$(RO)_3Si{-}(CH_2)_3{-}OOC(CH_3)C{=}CH_2$	Methacryl	MPS	Polyethylene Polyester
$(RO)_3Si{-}(CH_2)_3{-}SH$	Mercapto	MKPS	Natural rubber PVC
$(RO)_3Si{-}(CH_2)_3{-}O{-}CH_2CHCH_2O$	Glycidoxy	GPS	Epoxy Butyl rubber Polysulfide
$R_2{-}Si{-}Cl_2$	Chlorine	DCS	Polyethylene PVC
VTS grafted plastics	Vinyl	VSPP VSPE	Polypropylene Polyethylene
$(RO)_3{-}Si{-}R''{-}N_3$[b]	Azide	ATS	Polypropylene Polyethylene Polystyrene
$(RO)_3Si{-}(CH_2)_{15}CH_3$	Alkyl	HDS	Polyethylene Natural rubber

[a] $R = -$methyl or ethyl.
[b] $R'' = -C_6H_4{-}SO_2{-}$.

There are several reports indicating there is mechanical benefit from MA-based grafting. For example, Yu and co-workers [66] report an increase in σ_f from 105.2 MPa to 112.4 MPa in ramie fibre/PLA composites with a 3% MA addition. Using short pineapple leaf fibre and MA-g-PP in a polypropylene matrix, Hujuri and co-workers [67] note a ca. 17 MPa rise in σ_f and a ca. 660 MPa improvement in E_f relative to uncoupled fibres, at a fibre volume fraction of 10%. The same composites

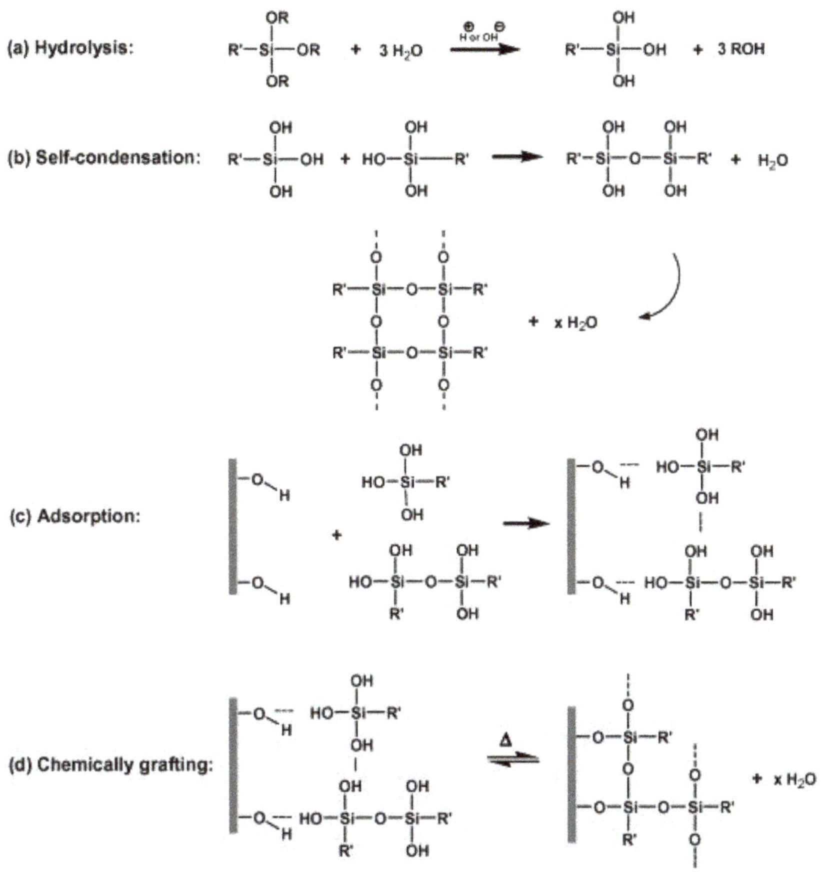

Figure 5.10. Interaction of silane coupling agents with natural fibres by (a) hydrolysis, (b) self-condensation, (c) adsorption, and (d) chemical grafting. Reprinted from [64]. Copyright (2010), with permission from Elsevier.

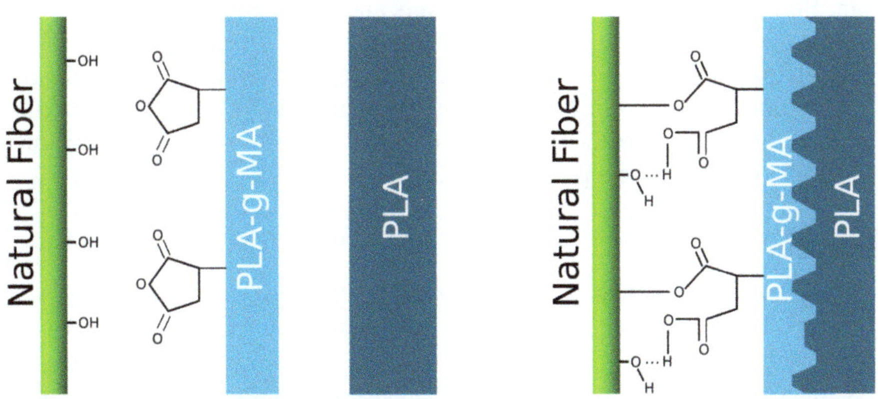

Figure 5.11. Interaction of PLA-g-MA coupling agents with natural fibres. Reproduced from [65]. CC BY 4.0.

subjected to impact loading revealed a 1.5-fold increase in the impact strength of MA-g-PP treated fibres, relative to untreated equivalents. The extent of MA-g-PP loading on the fibres has been shown to result in strength tensile strength improvement between loading concentrations of 0%–20% MA-g-PP [68], and it is theorised that as the depth of an MA-based sizing increases, so too does the extent to which it is able to mechanically entangle at the molecular level with the matrix polymer [69, 70].

5.4.5 Plasma treatment

Plasma treatment is essentially a process using ionised gas in a vacuum chamber to form plasma, which alters the surface of a material. The plasma treatment of natural fibres has also been shown to alter surface roughness and morphology. For example, Kafi and co-workers [71] report the RMS and R_a roughness of plasma-treated unwoven jute fibre fabric (Bangladesh Jute Mills Corporation, Dhaka) using three different gaseous mixtures. While the untreated jute fibre had $RMS = 41 \pm 19$ and $R_a = 32 \pm 12$, fibres treated using helium gas at a flow rate of 14 $\mathrm{l\,min^{-1}}$ measured $RMS = 15 \pm 6$ and $R_a = 12 \pm 4$; fibre treated with a gaseous mixture of helium and acetylene at flow rates of 14 $\mathrm{l\,min^{-1}}$ and 0.7 $\mathrm{l\,min^{-1}}$, respectively, measured $RMS = 28 \pm 10$ and $R_a = 22 \pm 11$; and fibres treated with a gaseous mixture of helium and nitrogen at flow rates of 14 $\mathrm{l\,min^{-1}}$ and 0.7 $\mathrm{l\,min^{-1}}$, respectively, measured $RMS = 46 \pm 13$ and $R_a = 34 \pm 9$. While Lyocell is not a purely natural fibre but rather a combination fibre of primarily dry jet-wet spun regenerated cellulose, it is included here as it is predominantly natural material based. Lyocell essentially is made using wood pulp together with N-methylmorpholine-N-oxide (an amine oxide), which precipitates into regenerated cellulose. Ercegovic and co-workers [72] discuss the effects of plasma treatment on the surface roughness of Lyocell, reporting an untreated fibre roughness of 122nm; oxygen plasma-treated Lyocell at 24 nm; argon plasma-treated Lyocell with a roughness between 23 nm and 59 nm; $AgNO_3$-treated fibres between 23 nm and 35 nm; and AgCl-treated fibres showing hierarchical roughness with the lower bound between 22 and 32 nm and the upper bound between 118 and 208 nm. Both plasma treatment output powers and treatment times have been shown to increase the topographical roughness of natural fibres. For example, Liu and Cheng [73] report that while untreated ramie fibres have on average a 35 nm roughness, plasma treatment at 100 w for 1 min increases roughness to 44.9 nm, plasma treatment at 150 w for 2 min increases the roughness to 47.8 nm, and a 3 min plasma treatment at 200 w results in a roughness profile of 48.3 nm.

References

[1] ASTM D2344/D2344M-22 *Standard Test Method for Short-Beam Strength of Polymer Matrix Composite Materials and Their Laminates* (West Conshohocken, PA: ASTM International)

[2] Benethuiliere T, Duchet-Rumeau J, Dubost E, Peyre C and Gerard J F 2020 Vinylester/glass fiber interface: still a key component for designing new styrene-free SMC composite materials *Compos. Sci. Technol.* **190** 108037

[3] Hansen C M 2007 *Hansen Solubility Parameters A User's Handbook* 2nd edn (Boca Raton, FL: CRC Press)
[4] van Krevelen D W and Nijenhuis K T 2009 *Properties of Polymer: Their Correlation with Chemical Structure; Their Numerical Estimation and Prediction from Additive Group Contributions* (Oxford: Elsevier)
[5] National Center for Biotechnology Information 2024 PubChem Compound Summary for CID 2286, Bisphenol A diglycidyl ether. Retrieved June 30, 2024 from https://pubchem.ncbi.nlm.nih.gov/compound/Bisphenol-A-diglycidyl-ether
[6] Diao F, Zhang Y, Liu Y, Fang J and Luan W 2016 γ-Ray irradiation stability and damage mechanism of glycidyl amine epoxy resin *Nucl. Inst. Methods Phys. Res. Section B: Beam Interact. Mater. Atoms* **383** 227–33
[7] George G A, Cole-Clarke P, St. John N and Friend G 1991 Real-time monitoring of the cure reaction of a TGDDM/DDS epoxy resin using fiber optic FT-IR *J. Appl. Polym. Sci.* **42** 643–57
[8] Ge H, Ma X and Liu H 2015 Preparation of emulsion-type thermotolerant sizing agent for carbon fiber and the interfacial properties of carbon fiber/epoxy resin composite *J. Appl. Polym. Sci.* **132** 41882
[9] Miura K and Nakata T 1986 Polyether polymer or copolymer, monomer therefor, and process for production thereof European Patent Office, Patent Number: EP0222586A2
[10] Gamstedt E K, Berglund L A and Peijs T 1999 Fatigue mechanisms in unidirectional glass-fibre-reinforced polypropylene *Compos. Sci. Technol.* **59** 759–68
[11] Oever M V D and Peijs T 1998 Continuous-glass-fibre-reinforced polypropylene composites. II. Influence of maleic-anhydride modified polypropylene on fatigue behaviour *Composites, Part* A **29** 227–39
[12] Oever M J A and Bos H L 1998 Critical fibre length and apparent interfacial shear strength of single flax fibre polypropylene composites *Adv. Compos. Lett.* **7** 81–5
[13] Gamstedt E K, Skrifvars M, Jacobsen T K and Pyrz R 2002 Synthesis of unsaturated polyesters for improved interfacial strength in carbon fibre composites *Composites A* **33** 1239–52
[14] Geng Y, Wang X, Yao J, Niu K and Yang C 2024 Preparation of unsaturated self-emulsifying polyester sizing agents for improving interfacial and mechanical properties of carbon fiber/vinyl ester resin composites *Composites A* **181** 108148
[15] Unterweger C, Duchoslav J, Stifter D and Furst C 2015 Characterization of carbon fiber surfaces and their impact on the mechanical properties of short carbon fiber reinforced polypropylene composites *Compos. Sci. Technol.* **108** 41–7
[16] Harper L T, Burn D T, Johnson M S and Warrior N A 2018 Long discontinuous carbon fibre/polypropylene composites for high volume structural applications *J. Compos. Mater.* **52** 1155–70
[17] Ozkan C, Karsli N G, Aytac A and Deniz V 2014 Short carbon fiber reinforced polycarbonate composites: effects of different sizing materials *Composites B* **62** 230–5
[18] Agopian J C, Teraube O, Dubois M and Charlet K 2020 Fluorination of carbon fibre sizing without mechanical or chemical loss of the fibre *Appl. Surf. Sci.* **534** 147647
[19] Li G, Zhang C, Wang Y, Li P, Yu Y, Jia X, Liu H, Yang X, Xue Z and Ryu S 2008 Interface correlation and toughness matching of phosphoric acid functionalized Kevlar fiber and epoxy matrix for filament winding composites *Compos. Sci. Technol.* **68** 3208–14

[20] Jiao Z, Yao Z, Zhou J, Yi P and Lu C 2021 Reinforced interface and mechanical properties of high strength carbon fiber composites *High Perform. Polym.* **33** 255–63
[21] Yang T, Zhao Y, Liu H, Sun M and Xiong S 2021 Effect of sizing agents on surface properties of carbon fibers and interfacial adhesion of carbon fiber/bismaleimide composites *ACS Omega* **6** 23028–37
[22] Vautard F and Drzal L T 2009 Carbon fibre vinyl ester interfacial adhesion by the use of a reactive epoxy coating *Proc. 17th Int. Committee on Composite Materials ICCM-17* (Edinburgh, Scotland, 27 July–31 2009)
[23] Alam P 2021 *Composites Engineering: An A-Z Guide* (Bristol: IOP Publishing)
[24] Thomason J L 2012 *Glass Fibre Sizings: A Review of the Scientific Literature* (Amsterdam: Elsevier)
[25] National Center for Biotechnology Information 2024 PubChem Compound Summary for CID 129 660 742, gamma-(Aminopropyl)triethoxysilane Retrieved June 28, 2024 from https://pubchem.ncbi.nlm.nih.gov/compound/gamma-_Aminopropyl_triethoxysilane
[26] National Center for Biotechnology Information 2024 PubChem Compound Summary for CID 17 317, (3-Glycidoxypropyl)trimethoxysilane Retrieved June 28, 2024 from https://pubchem.ncbi.nlm.nih.gov/compound/3-Glycidoxypropyl_trimethoxysilane
[27] ChemSpider (Royal Society of Chemistry), 3-(Trimethoxysilyl)propyl methacrylate, CSID:16388, http://www.chemspider.com/Chemical-Structure.16388.html [accessed 17:12, Jun 28, 2024]
[28] National Center for Biotechnology Information 2024 PubChem Compound Summary for CID 6516, Vinyltriethoxysilane Retrieved June 28, 2024 from https://pubchem.ncbi.nlm.nih.gov/compound/Vinyltriethoxysilane
[29] Thomason J L 2015 *Glass Fibre Sizing: A Review of Size Formulation Patents* (Amsterdam: Elsevier)
[30] Yang X, Zhou S, Zhang X, Xiang L, Xie B and Luo X 2022 Enhancing oxygen/moisture resistance of quantum dots by short-chain, densely cross-linked silica glass network *Nanotechnology* **33** 465202
[31] Thomason J L 2019 Glass fibre sizing: a review *Composites A* **127** 105619
[32] Arunprakash V R and Rajadurai A 2017 Inter laminar shear strength behavior of acid, base and silane treated E-glass fibre epoxy resin composites on drilling process *Defence Technol.* **13** 40–6
[33] Hamada H, Fujihara K and Harada A 2000 The influence of sizing conditions on bending properties of continuous glass fiber reinforced polypropylene composites *Composites A* **31** 979–90
[34] Jeevi G, Ranganathan N and Abdul Kader M 2022 Studies on mechanical and fracture properties of basalt/E-glass fiber reinforced vinyl ester hybrid composites *Polym. Compos.* **43** 3609–25
[35] Kiss P, Schoefer J, Stadlbauer W, Burgstaller C and Archodoulaki V M 2021 An experimental study of glass fibre roving sizings and yarn finishes in high-performance GF-PA6 and GF-PPS composite laminates *Composites B* **204** 108487
[36] Mader E, Jacobasch H J, Grunke K and Gietzelt T 1996 Influence of an optimized interphase on the properties of polypropylene/glass fibre composites *Composites A* **27** 907–12
[37] Dey M, Deitzel J M, Gillespie J W Jr and Schweiger S 2014 Influence of sizing formulations on glass/epoxy interphase properties *Composites A* **63** 59–67

[38] Alam P, Robert C and O Bradaigh C M 2018 Tidal turbine blade composites–a review on the effects of hygrothermal aging on the properties of CFR *Composites B* **149** 248–59
[39] Shahzad A 2012 Effects of alkalization on tensile, impact, and fatigue properties of hemp fiber composites *Polym. Compos.* **33** 1129–40
[40] Chand N and Fahim M 2008 Natural fibers and their composites *Woodhead Publishing Series in Composites Science and Engineering, Tribology of Natural Fiber Polymer Composites* (Cambridge: Woodhead Publishing Ltd)
[41] Widodo E, Pratikto , Sugiarto and Widodo T D 2024 Comprehensive investigation of raw and NaOH alkalized *sansevieria* fiber for enhancing composite reinforcement *Case Studies Chem. Environ. Eng.* **9** 100546
[42] Mwaikambo L 2009 Tensile properties of alkalised jute fibres *Bioresources* **4** 566–88
[43] Senthamaraikannan P and Kathiresan M 2018 Characterization of raw and alkali treated new natural cellulosic fiber from *Coccinia grandis.L Carbohydrate Polym.* **186** 332–43
[44] Verma D and Goh K L 2021 Effect of mercerization/alkali surface treatment of natural fibres and their utilization in polymer composites: mechanical and morphological studies *J. Compos. Sci.* **5** 175
[45] Mwaikambo L and Ansell M P 2003 Hemp fibre reinforced cashew nut shell liquid composites *Compos. Sci. Technol.* **63** 1297–305
[46] Ouajai S, Hodzic A and Shanks R A 2004 Morphological and grafting modification of natural cellulose fibers *J. Appl. Polym. Sci.* **94** 2456–65
[47] Pickering K L, Beckermann G W, Alam S N and Foreman N J 2007 Optimising industrial hemp fibre for composites *Composites A* **38** 461–8
[48] Kostic M, Pejic B and Skundric P 2008 Quality of chemically modified hemp fibers *Bioresour. Technol.* **99** 94–9
[49] Mwaikambo L Y and Ansell M P 2006 Mechanical properties of alkali treated plant fibres and their potential as reinforcement materials. I. hemp fibres *J. Mater. Sci.* **41** 2483–96
[50] Mwaikambo L Y and Ansell M P 2006 Mechanical properties of alkali treated plant fibres and their potential as reinforcement materials. II. sisal fibres *J. Mater. Sci.* **41** 2497–508
[51] Raju J S N, Depoures M V and Kumaran P 2021 Comprehensive characterization of raw and alkali (NaOH) treated natural fibers from Symphirema involucratum stem *Int. J. Biol. Macromol.* **186** 886–96
[52] Rokbi M, Osmani H, Imad A and Benseddiq N 2011 Effect of chemical treatment on flexure properties of natural fiber-reinforced polyester composite *Procedia Eng.* **10** 2092–7
[53] Chikouche M D L, Merrouche A, Azizi A, Rokbi M and Walter S 2015 Influence of alkali treatment on the mechanical properties of new cane fibre/polyester composites *J. Reinf. Plast. Compos.* **34** 1329–39
[54] Cavalcanti D K K, Banea M D, Neto J S S, Lima R A A, da Silva L F M and Carbas R J C 2019 Mechanical characterization of intralaminar natural fibre-reinforced hybrid composites *Composites B* **175** 107149
[55] Jamilah U L and Sujito S 2021 The improvement of ramie fiber properties as composite materials using alkalization treatment: NaOH concentration *J. Mater. Sci.* **22** 62–70
[56] Manalo A C, Wani E, Zukarnain N A, Karunasena W and Lau K T 2015 Effects of alkali treatment and elevated temperature on the mechanical properties of bamboo fibre-polyester composites *Composites B* **80** 73–83

[57] de Weyenberg I V, Truong T C, Vangrimde B and Verpoest I 2006 Improving the properties of UD flax fibre reinforced composites by applying an alkaline fibre treatment *Composites A* **37** 1368–76

[58] George M, Mussone P G, Abboud Z and Bressler D C 2014 Characterization of chemically and enzymatically treated hemp fibres using atomic force microscopy and spectroscopy *Appl. Surf. Sci.* **314** 1019–25

[59] Tserki V, Panayiotou C and Zafeiropoulos N E 2005 A study of the effect of acetylation and propionylation on the interface of natural fibre biodegradable composites *Adv. Compos. Lett.* **14**

[60] Ozemn N 2012 A study of the effect of acetylation on hemp fibres with vinyl acetate *Bioresources* **7** 3800–9

[61] Sreekala M S and Thomas S 2003 Effect of fibre surface modification on water-sorption characteristics of oil palm fibres *Compos. Sci. Technol.* **63** 861–9

[62] Oladele I O, Michael O S, Adediran A A, Balogun O P and Ajagbe F O 2020 Acetylation treatment for the batch processing of natural fibers: effects on constituents, tensile properties and surface morphology of selected plant stem fibers *Fibers* **8** 73

[63] Khalil H P S A, Ismail H, Rozman H D and Ahmad M N 2001 The effect of acetylation on interfacial shear strength between plant fibres and various matrices *Eur. Polym. J.* **37** 1037–45

[64] Xie Y, Hill C A S, Xiao Z, Militz H and Mai C 2010 Silane coupling agents used for natural fiber/polymer composites: a review *Composites A* **41** 806–19

[65] Oliver-Ortega H, Reixach R, Espinach F X and Méndez J A 2022 Maleic anhydride polylactic acid coupling agent prepared from solvent reaction: synthesis, characterization and composite performance *Materials* **15** 1161

[66] Yu T, Jiang N and Li Y 2014 Study on short ramie fiber/poly(lactic acid) composites compatibilized by maleic anhydride *Composites A* **64** 139–46

[67] Hujuri U, Chattopadhay S K, Uppulari R and Ghoshal A K 2008 Effect of maleic anhydride grafted polypropylene on the mechanical and morphological properties of chemically modified short-pineapple-leaf-fiber-reinforced polypropylene composites *J. Appl. Polym. Sci.* **107** 1507–16

[68] Techawinyutham L, Frick A and Siengchin S 2016 Polypropylene/maleic anhydride grafted polypropylene (MAgPP)/coconut fiber composites *Adv. Mech. Eng.* **8**

[69] Franco-Marques E, Mendez J, Pelach M, Vilaseca F, Bayer J and Mutje P 2011 Influence of coupling agents in the preparation of polypropylene composites reinforced with recycled fibers *Chem. Eng. J.* **166** 1170–8

[70] Anbupalani M S, Venkatachalam C D and Rathanasamy R 2020 Influence of coupling agent on altering the reinforcing efficiency of natural fibre-incorporated polymers–a review *J. Reinf. Plast. Compos.* **39** 520–44

[71] Kafi A A, Magniez K and Fox B L 2011 A surface-property relationship of atmospheric plasma treated jute composites *Compos. Sci. Technol.* **71** 1692–8

[72] Ercegovic S R, Cunko R, Svetlicic V and Segota S 2011 Application of AFM for identification of fibre surface changes after plasma treatments *Mater. Technol.* **26** 146–52

[73] Liu X and Cheng L 2016 Influence of plasma treatment on properties of ramie fiber and the reinforced composites *J. Adhes. Sci. Technol.* **31** 1723–34

IOP Publishing

Composite Interfaces in Mechanical Design

Parvez Alam

Chapter 6

Interface morphology

6.1 Introduction

Surface morphology is a generic term that covers all aspects of the shape and form of surfaces. Within composites engineering, the majority of research efforts on morphology are simplified mathematical descriptions of surface roughness. As such, much of this chapter will be dedicated to understanding surface morphology in terms of roughness and mechanical properties. Fibre surface morphologies naturally differ based on the type of material used, the manufacturing or processing method, the fibre source, and the surface treatment or finish. In this chapter, a few fundamental aspects of roughness and morphology will be discussed, highlighting certain reports from the literature for different reinforcing fibres, more specifically: glass fibre and carbon fibre. The effects roughness may have on the mechanical and interfacial properties of glass and carbon fibre-reinforced composites will also be detailed.

6.2 Types of roughness

While there are many roughness parameters, composites engineers have tended to use only a few to characterise the surface roughness of fibre reinforcements. The most commonly used is by far the average roughness, R_a, followed by the root mean squared, *RMS,* roughness, an equivalent to which is the root mean square deviation roughness, R_q. In addition to these, there are fewer reports using the maximum peak to valley height roughness, R_z. These roughness parameters are popular because they are easily measurable by atomic force microscopy (AFM), optical and 3D profilometry, confocal and scanning electron microscopy, etc. Modern ISO standards related to surface roughness are provided in table 6.1, many of which replace, or will shortly replace, older standards (as indicated in the table).

The R_a roughness is essentially a measure of average height, figure 6.1(a). To deduce the 'height' a mean line, *m*, is set asymptotically with a chosen *x*-Cartesian as a best fit between the maximum and minimum peaks and troughs. Heights are

doi:10.1088/978-0-7503-5688-6ch6

Table 6.1. Current ISO standards related to the characterisation of surface roughness and the standards they replace.

ISO standard number and name	Source
ISO 21920-1:2021 Geometrical product specifications (GPS)—Surface texture: Profile Part 1: Indication of surface texture	[1]
ISO 21920-2:2021 Geometrical product specifications (GPS)—Surface texture: Profile Part 2: Terms, definitions and surface texture parameters (replaces: ISO 4287:1997 Geometrical Product Specifications (GPS)—Surface texture: Profile method—Terms, definitions and surface texture parameters)	[2]
ISO 21920-3:2021 Geometrical product specifications (GPS)—Surface texture: Profile Part 3: Specification operators (replaces: ISO 4288:1996 Geometrical Product Specifications (GPS—Surface texture: Profile method—Rules and procedures for the assessment of surface texture)	[3]
ISO 8785:1998 Geometrical Product Specification (GPS)—Surface imperfections—Terms, definitions and parameters	[4]
ISO 16610-21:2011 Geometrical product specifications (GPS)—Filtration Part 21: Linear profile filters: Gaussian filters (to be replaced by: ISO/DIS 16610-21 Geometrical product specifications (GPS)—Filtration Part 21: Linear profile filters: Gaussian filters)	[5]

measured from m at each individual peak and trough point of inflection along the x-Cartesian distance from $0 \rightarrow l$, where l is the sampling distance of x along which measurements are being taken. The peak and trough heights are taken as absolute values and R_a is thence calculated according to equation (6.1).

$$R_a = \frac{1}{l}\int_0^l |y(x)|\mathrm{d}x \tag{6.1}$$

The *RMS* roughness takes y-Cartesian distances at intervals within the peaks and troughs over the sampling distance, l, over which measurements are taken from sample counts i where $i = 1 \rightarrow n$, equation (6.2). Figure 6.1(b) provides an illustrative representation of the measurements and parameters.

$$RMS = R_q = \sqrt{\frac{1}{n}\sum_{i=1}^{n} y_i^2} \tag{6.2}$$

The R_z roughness takes y-Cartesian distances of a specified number of sample counts, i, using the highest peaks and troughs over the sampling distance, l, where

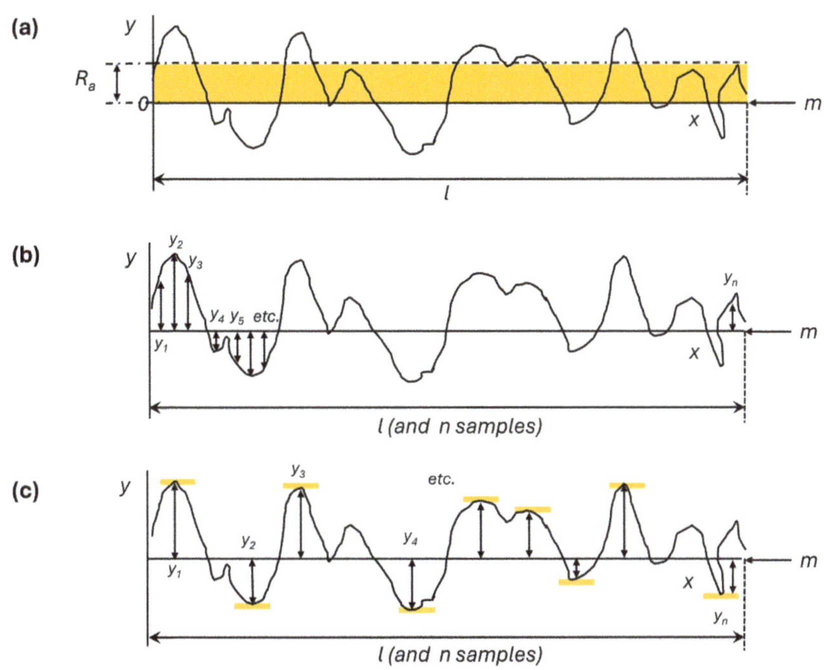

Figure 6.1. Illustrative representations of (a) R_a roughness, (b) *RMS* roughness, and (c) R_z roughness.

$i = 1 \rightarrow n$, equation (6.3). Figure 6.1(c) provides an illustrative representation of the measurements and parameters.

$$R_z = \frac{1}{n}\sum_{i=1}^{n}|y_i| \tag{6.3}$$

6.3 Glass fibre roughness, morphology, and properties

Glass fibres are common engineering fibres applied as reinforcing within polymers such as epoxides and unsaturated polyesters. They are essentially non-crystalline (amorphous) isotropic solids, with no distinct microstructure. Common types of glass fibres used in composites are E-glass (electronic glass) and S-glass (high silica content glass). In general, the surface treatment of glass fibre, primarily by application of sizing, increases its roughness. This can be evidenced from a number of studies, table 6.2, a few of which will be discussed here in terms of roughness and its effects on interface properties. Gao and co-workers [6] (table 6.2 first band) considered the effect of sizing on the roughness and morphology of E-glass. In their work they applied a range of different sizing formulations including 3-glycidyloxypropylmethylsilane (GPS), propyltrimethoxysilane (PTMO), and bisphenol A diglycidyl ether (DGEBA) film formers. When comparing against unsized equivalents ($R_a = 5.18$ nm ± 1.56 nm), they report the R_a roughness being highest on fibres sized with GPS/PTMO silanes, DGEBA epoxy film former, and colloidal silica roughening agent (18.01 nm ± 6.86 nm), followed by fibres sized only with

Table 6.2. Example roughness values for a range of glass fibres. Here: GPS is 3-glycidyloxypropylmethylsilane, PTMO is propyltrimethoxysilane, DGEBA is bisphenol A diglycidyl ether, γ-APS is γ-aminopropyltriethoxy, PU is polyurethane, and PP is polypropylene.

Roughness	Parameter	GLASS FIBRE Details
5.18 nm ± 1.56 nm	R_a	E-glass fibre (diameter 18 μm) from Fiber Glass Industries, Inc. (unsized) [6]
18.01 nm ± 6.86 nm	R_a	E-glass fibre (diameter 18 μm) from Fiber Glass Industries, Inc. sized with GPS/PTMO silanes, DGEBA epoxy film former (dispersion), and colloidal silica roughening agent [6]
7.89 nm ± 3.49 nm	R_a	E-glass fibre (diameter 18 μm) from Fiber Glass Industries, Inc. sized with GPS silane and DGEBA epoxy film former (dispersion) [6]
11.74 nm ± 3.91 nm	R_a	E-glass fibre (diameter 18 μm) from Fiber Glass Industries, Inc. sized with GPS/PTMO silanes and DGEBA epoxy film former (dispersion) [6]
5.48 nm ± 1.28 nm	R_a	E-glass fibre (diameter 18 μm) from Fiber Glass Industries, Inc. sized with PTMO silane and DGEBA epoxy film former (dispersion) [6]
5.4 nm	R_a	E-glass fibre (diameter of 13–31 μm) manufactured at the Institute of Polymer Research, Dresden (unsized) [7]
8.3 nm	R_a	E-glass fibre (diameter of 13–31 μm) manufactured at the Institute of Polymer Research, Dresden sized with γ-APS silane and PU film former [7]
15 nm	R_a	E-glass fibre (diameter of 13–31 μm) manufactured at the Institute of Polymer Research, Dresden sized with γ-APS silane and PP film former [7]
0.8 nm ± 0.5 nm	R_q	Glass fibre (diameter of 15 μm) manufactured at the Leibniz Institute of Polymer Research, Dresden sized with γ-APS silane [8]
2.1 ± 1.1 nm ± 0.5 nm	R_q	Glass fibre (diameter of 15 μm) manufactured at the Leibniz Institute of Polymer Research, Dresden sized with PP dispersion [8]
16.1 nm ± 7.2 nm	R_q	Glass fibre (diameter of 15 μm) manufactured at the Leibniz Institute of Polymer Research, Dresden sized with γ-APS silane (first stage) and PP dispersion (second stage) [8]
18.6 nm ± 2.7 nm	R_q	Glass fibre (diameter of 15 μm) manufactured at the Leibniz Institute of Polymer Research, Dresden sized with γ-APS silane (first stage) and PP dispersion (second stage) using an increased applicator roll speed [8]
4.9 nm ± 1.5 nm	R_q	Glass fibre (diameter of 15 μm) manufactured at the Leibniz Institute of Polymer Research, Dresden sized with PP dispersion (first stage) and γ-APS silane (second stage) [8]

5.8 nm ± 1.8 nm	R_q	Glass fibre (diameter of 15 μm) manufactured at the Leibniz Institute of Polymer Research, Dresden sized with γ-APS silane and PP dispersion (both within the first stage) [8]
11 nm ± 4 nm	R_a	S-glass fibre (diameter 9–12 μm) from Owens Corning sized with γ-GPS silane, and DGEBA epoxy film former [9]
13 nm ± 5 nm	R_a	S-glass fibre (diameter 9–12 μm) from Owens Corning sized with γ-GPS/γ-PTMO silanes, and DGEBA epoxy film former [9]
5 nm ± 2 nm	R_a	S-glass fibre (diameter 9–12 μm) from Owens Corning sized with γ-APS silane, and silylated polyazamide film former [9]
15 nm ± 6 nm	R_a	S-glass fibre (diameter 9–12 μm) from Owens Corning sized with γ-APS silane, and hydrosize polyurethane film former [9]
5 nm ± 3 nm	R_a	S-glass fibre (diameter 9–12 μm) from Owens Corning sized with γ-GPS/γ-APS/γ-PTMO silanes, and epoxy + hydrosize polyurethane film former [9]

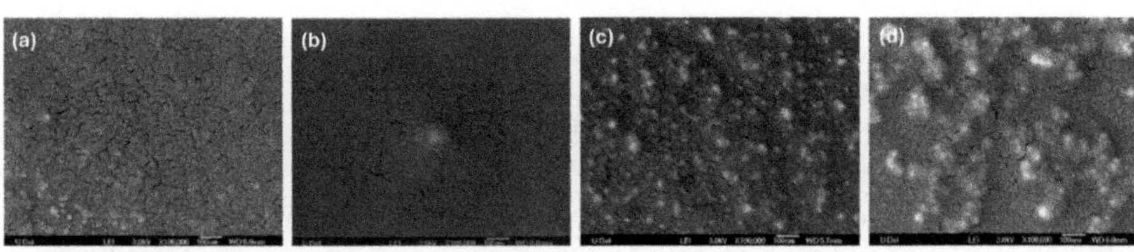

Figure 6.2. SEM micrographs showing the roughness profiles of (a) PTMO silane only with DGEBA epoxy film former, (b) GPS silane only with DGEBA epoxy film former, (c) GPS/PTMO silanes and DGEBA epoxy film former, and (d) GPS/PTMO silanes, DGEBA epoxy film former, and colloidal silica roughening agent. Reprinted from [6]. Copyright (2015), with permission from Elsevier.

GPS/PTMO silanes and DGEBA epoxy film former and without the additional roughening agent (11.74 nm ± 3.91 nm). In both cases, a size coupling (GPS and PTMO) was used. Importantly, they note that the use of single silane sizings (i.e. either GPS or PTMO) resulted in lower levels of roughness than in the case of the coupled (GPS/PTMO) applied sizing. These include GPS silane only with DGEBA epoxy film former (7.89 nm ± 3.49 nm) and PTMO silane only with DGEBA epoxy film former (5.48 nm ± 1.28 nm). The different surfaces are shown in figure 6.2. A point of interest and note is that when conducting microbond tests, Gao and co-workers [6] report a clear correlation between the surface roughness of a fibre and both its energy of debond, E_{debond}, from the resin droplet, and its subsequent sliding energy, $E_{sliding}$, with the higher roughness fibres requiring more of both E_{debond} and $E_{sliding}$ than fibres with lower roughness, figure 6.3(a). This trend is generally also true when the same fibres are used to reinforce polymer and tested for interfacial

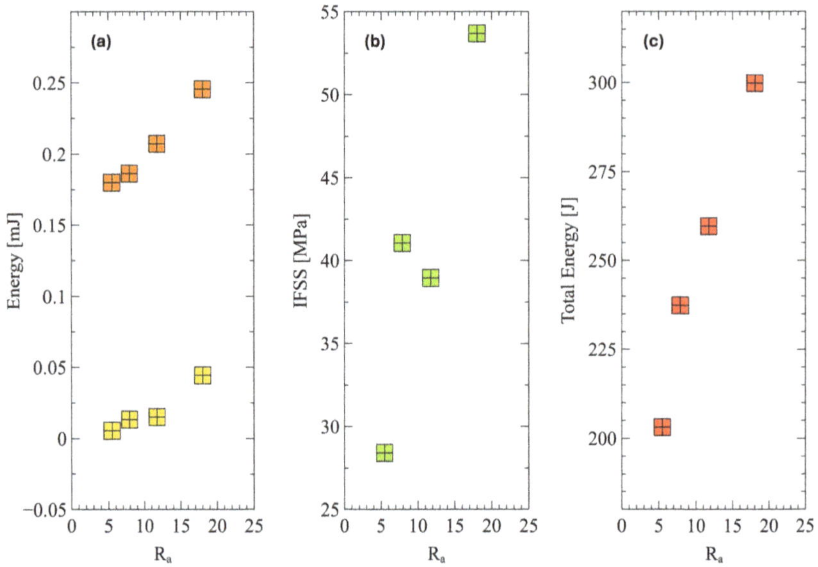

Figure 6.3. Plots made using Gao and co-workers' data [6] showing the potential effects of R_a roughness on (a) debonding (yellow) and sliding (orange) energies, (b) IFSS, and (c) the total energy measured from IFSS and punch shear testing.

shear strength (IFSS), figure 6.3(b), and in terms of the total energy output from IFSS testing and punch shear testing, figure 6.3(c).

While in the work of Gao *et al* [6], DGEBA epoxy was used consistently as a film former, varying the film former can also affect the final roughness of sized glass. In a study by Gao and Mader, [7] (table 6.2 second band), γ-aminopropyltriethoxy silane (γ-APS silane) was used consistently to size E-glass, whilst varying the film former (using either polyurethane, PU, or polypropylene, PP). In their paper they note that the use of PP film former almost doubles the R_a roughness (15 nm) as compared to when PU film former is used (8.3 nm). These noteworthy differences in surface roughness appear to correlate with mechanical performance. Comparing the properties of 13.4% volume fraction short fibre-reinforced PP matrix GFRP composites using fibres of each type, fibre pullout strength, τ_p, and notched Charpy impact toughness, *CVN*, is ca. 2-fold and 3-fold higher, respectively, in γ-APS/PP-sized GFRP composites ($\tau_p = 21.7$ MPa, $CVN = 35$ kJm^{-2}) than in γ-APS/PU-sized GFRP composites ($\tau_p = 10.9$ MPa, $CVN = 13.3$ kJ m^{-2}). Similarly, static properties such as strength, σ, and elastic modulus, E, were higher in γ-APS/PP-sized GFRP composites ($\sigma = 73.6$ MPa, $E = 7.2$ GPa) than in γ-APS/PU-sized GFRP composites ($\sigma = 37.4$ MPa, $E = 6.2$ GPa). Similar correlations were reported by Gao and Mader [7] for 50% volume fraction unidirectional composites in epoxy matrix such that the ultimate shear strengths, τ_{ult}, reported here were 84.3 MPa in γ-APS/PU-sized GFRP composites and 59.7 in unsized-GFRP composites, while static properties included $\sigma = 49.2$ MPa and $E = 9.31$ MPa in γ-APS/PU-sized GFRP composites while in unsized equivalents, $\sigma = 27.3$ MPa and $E = 8.69$ MPa.

The order in which surface treatments are applied can also affect the final roughness of a surface-treated glass fibre. This phenomenon has been documented, for example, by Zhuang and co-workers [8] who varied the stages at which γ-APS silane and PP dispersion were used during the sizing process (see table 6.2 third band). They note, for example, that applying γ-APS silane during the first stage, followed by PP in a second stage of processing, resulted in a ca. 3-fold higher R_q roughness (16.1 nm ± 7.2 nm) than if applying both γ-APS silane and PP within the first stage (5.8 nm ± 1.8 nm). In addition, an increase in the applicator roll speed is reported to further raise the roughness of two-stage γ-APS silane/PP from 16.1 nm ± 7.2nm to 18.6 nm ± 2.7 nm. The work of Dey and co-workers [9] summarises several of the concepts discussed here (see table 6.2 fourth band), including differences arising from the use of either single or combined silane sizings and differences arising from variations in the film former. Despite the evident parallels in regards to roughness and topography, Dey an co-workers' outputs differ from previously discussed works. While Dey *et al* report a positive correlation between sliding energy and surface roughness, figure 6.4(a), there is in fact a negative correlation between IFSS and R_a roughness, figure 6.4(b), which is in contrast to the positive correlation between IFSS and R_a roughness as noted from

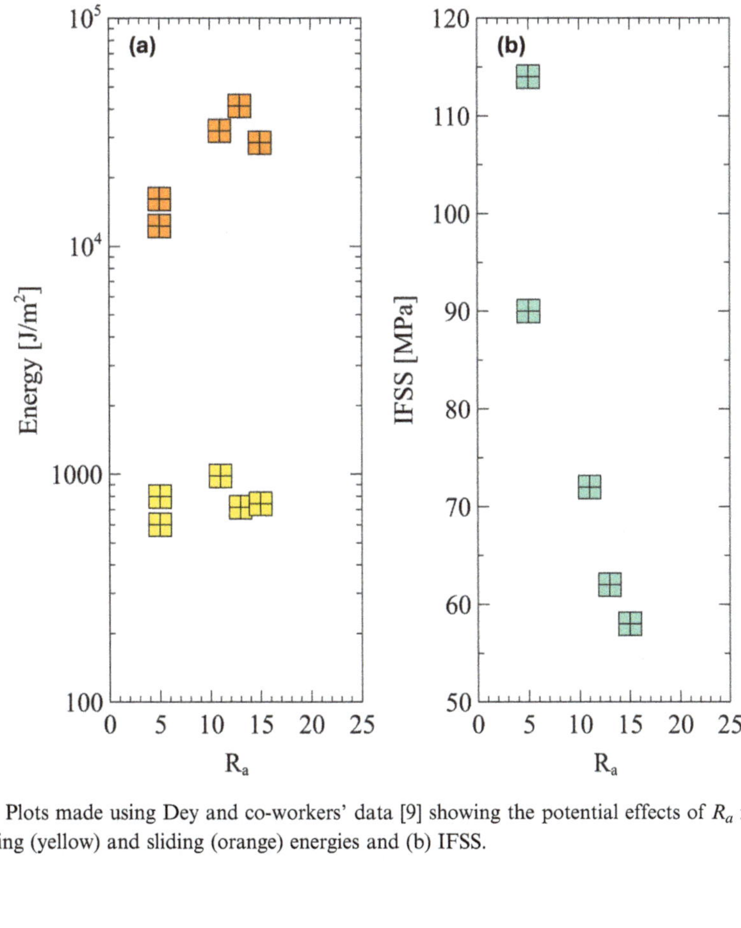

Figure 6.4. Plots made using Dey and co-workers' data [9] showing the potential effects of R_a roughness on (a) debonding (yellow) and sliding (orange) energies and (b) IFSS.

Gao's work [6] plotted in figure 6.3. The expectation is that as roughness increases, so too does the surface area and the potential for mechanical interlocking of adhesion. The work of Rey and co-workers highlights an important aspect of interfacial adhesion and roughness, as roughness *alone* cannot define the strength of adhesion at fibre–matrix interfaces. First, roughness alone does not guarantee good wetting behaviour as discussed further in chapter 3. Additionally, the sizing materials used by Rey and co-workers are somewhat variable in terms of their chemistry and hence their potential interactions with matrix matter. This in turn may affect other mechanisms of adhesion including electrostatic, chemical, and diffusion-based mechanisms, each of which is described in more detail in chapter 3.

6.4 Carbon fibre roughness, morphology, and properties

Table 6.3 provides roughness values for a range of different carbon fibre types, processing parameters, and surface treatments. In the initial data set (shown within the first band [10]) R_a roughness values are determined by AFM (scope: 3 $\mu m \times$ 3 μm) from the surfaces of PAN-based T300 fibres (diameter 7 μm) (Jilin Chemical Industrial Company, China) that were initially acetone treated to remove both impurities and sizing agent, and then surface treated with aqueous ammonia (NH_3) and dried at 120 °C for 3 h. This combination of surface treatments results in both the erosion and the oxidation of the carbon fibre surface [11]. The erosive nature of NH_3 is evidenced by the increase in roughness as a function of prolonged NH_3 exposure [10]. An important concept can be derived from this work as increasing exposure time increases the sizing roughness as elucidated in the table. Since the size applied to the fibre expands the radial diameter of the fibre and the sizing has inferior mechanical properties, the overall fibre tensile strength is reduced, as shown in figure 6.5(a). Nevertheless, the size and its roughness (increasing as a function of treatment time) improves mechanical interlocking and broadens the surface area available for adhesion. As a result of this, the IFSS increases as a function of treatment time, as shown in figure 6.5(b).

The second band of data for polyacryonitrile-based 6K diameter carbon fibres (7 μm) (Weihai Guangwei Group Co. Ltd, China) shows high levels of variability in R_a surface roughness. In the study by Ruan and co-workers [12] three fibre types (CF-1, CF-2, and CF-3) were subject to acetone treatment after which they underwent sonification to aid in the removal of sizing. Even though the polyacryonitrile-based 6K diameter carbon fibres were desized in the same manner, the final unsized fibre showed significant morphological variability, figure 6.6. The roughness values varied from 1.6 nm (CF-1) to 42 nm (CF-2) to 78 nm (CF-3).

In the third band the R_a roughness of an intermediate modulus (IM) and two high modulus (HM) carbon fibres are provided as 'before' and 'after' single fibre pullout from an epoxy resin matrix. While there is a 10-fold difference in the R_a roughness of the IM relative to the HM fibres, there appears to be no knock-on effect with regard to the roughness of the epoxy layer attached to the surface of the fibre after pullout testing, as can be deduced from the overlapping standard deviations between the three groups [13]. The roughness may nevertheless have a knock-on effect on the

Table 6.3. Example roughness values for a range of carbon fibres. IM = intermediate modulus, HM = high modulus, PAN = polyacrylonitrile.

Roughness	Parameter	CARBON FIBRE Details
12.5 nm	R_a	PAN-based T300 (unsized) [10]
19.2 nm	R_a	PAN-based T300 (unsized) 24 h NH_3 treated [10]
25.2 nm	R_a	PAN-based T300 (unsized) 48 h NH_3 treated [10]
32.1 nm	R_a	PAN-based T300 (unsized) 72 h NH_3 treated [10]
42.3 nm	R_a	PAN-based T300 (unsized) 96 h NH_3 treated [10]
57.4 nm	R_a	PAN-based T300 (unsized) 120 h NH_3 treated [10]
1.6 nm	R_a	PAN-based 6 K type 1 (unsized) [12]
42 nm	R_a	PAN-based 6 K type 2 (unsized) [12]
78 nm	R_a	PAN-based 6 K type 3 (unsized) [12]
0.3 nm	R_a	IM (finished) [13]
4.2 nm $\pm 2.5 nm$	R_a	IM (finished) after pullout test from epoxy [13]
3.8 nm	R_a	HM (oxidised without finish) [13]
2.7 nm $\pm 1.1 nm$	R_a	HM (oxidised without finish) after pullout test from epoxy [13]
4.0 nm	R_a	HM (finished) [13]
3.7 nm $\pm 1.3 nm$	R_a	HM (finished) after pullout test from epoxy [13]
7.5 nm	R_a	PAN-based AS-4 (unsized) [14]
9.1 nm	RMS	PAN-based AS-4 (unsized) [14]
9.0 nm	R_a	PAN-based AS-4 (polyetherimide sized) [14]
11.0 nm	RMS	PAN-based AS-4 (polyetherimide sized) [14]
15.0 nm	R_a	PAN-based AS-4 (poly(thioarylene phosphine oxide) sized) [14]
18.0 nm	RMS	PAN-based AS-4 (poly(thioarylene phosphine oxide) sized) [14]
37 nm $\pm$ 17nm	R_a	Panex 35 PAN-based 50 K tow automotive grade fibres (carbonised (unoxidised)) [15]
50 nm $\pm$ 10nm	R_a	Panex 35 PAN-based 50 K tow automotive grade fibres (oxidised) [15]
33 nm $\pm$ 13nm	R_a	Panex 35 PAN-based 50 K tow automotive grade fibres (sized) [15]
1.45 nm	R_a	Magnamite® IM6 (unsized) [16]
0.78 nm	R_a	Magnamite® IM6 (20% of nominal sizing) [16]
0.60 nm	R_a	Magnamite® IM6 (100% of nominal sizing) [16]
1.08 nm	R_a	Magnamite® IM6 (200% of nominal sizing) [16]
2.00 nm	R_a	Magnamite® IM6 (600% of nominal sizing) [16]

interfacial properties of the fibres as both the interlaminar shear strength (ILSS) and the critical shear strain energy release rate, G_{IIc}, are higher in the lower R_a roughness IM fibres (ILSS = 84 ± 4 MPa, $G_{IIc} = 452 \pm 35$ Jm^{-2}) than in the HM fibres (ILSS = 75 ± 3 MPa, $G_{IIc} = 272 \pm 26$ $J\,m^{-2}$).

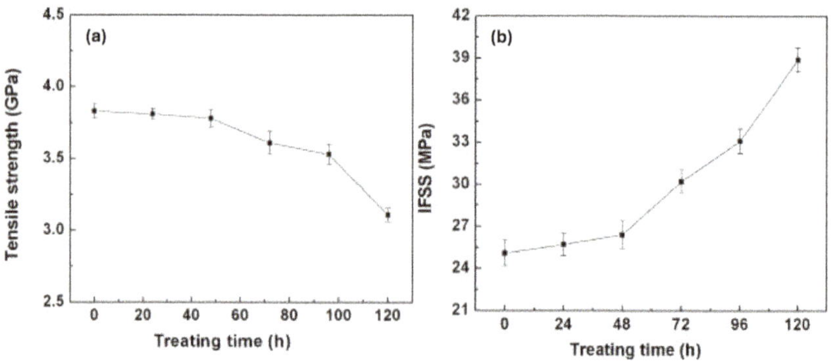

Figure 6.5. The effect of NH_3 treatment time on (a) the tensile strength of sized individual carbon fibres and (b) the IFSS of unsized and sized carbon fibre-reinforced epoxy composites (note: surface roughness increases as a function of NH_3 treatment time). Reprinted from [10]. Copyright (2011), with permission from Elsevier.

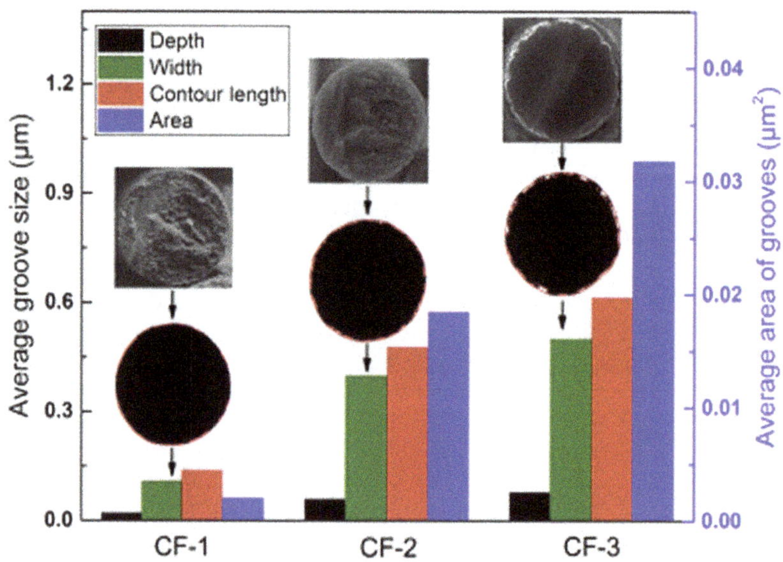

Figure 6.6. Morphological variations (groove size) measured in CF-1, CF-2, and CF-3 polyacryonitrile-based 6K diameter carbon fibres (7 μm) (Weihai Guangwei Group Co. Ltd, China) subjected to desizing via acetone treatment and sonification. Reprinted from [12]. CC BY NC-ND.

Data sets comparing Hercules polyacrylonitrile-based carbon fibers (AS-4) using both R_a and *RMS* roughness are represented in the fourth band of the table [14]. This data demonstrates how the roughness model impacts the final value of roughness recorded, with *RMS* roughness providing consistently higher values within the data set than the R_a roughness. Using both roughness models, polythioarylene phosphine oxide-sized fibres are most rough, followed by polyetherimide-sized fibres, followed by unsized carbon fibres. Surface roughness may vary during the different manufacturing stages of a finished fibre [15]. This is demonstrated in the fifth band of

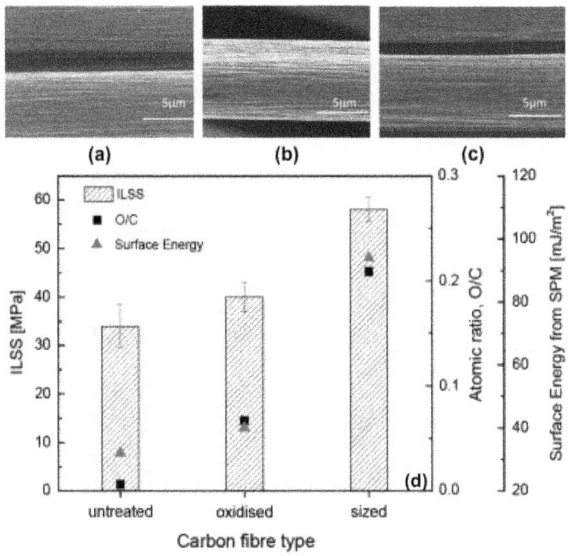

Figure 6.7. Morphological variations shown using SEM images for (a) unoxidised, (b) oxidised, and (c) sized carbon fibres; and (d) atomic ratio (O/C) and surface energy versus ILSS for unoxidised, oxidised, and epoxy sized carbon fibres. Reprinted from [15]. Copyright (2014), with permission from Elsevier.

table 6.3 where the R_a roughness of Panex 35 PAN-based 50 K tow automotive grade fibres supplied by Zoltek Companies, Inc., Hungary was compared at the fibre processing stages of carbonisation, oxidation, and post-sizing (using an epoxy size). Figures 6.7(a)–(c) show the surface morphologies of (a) unoxidised, (b) oxidised, and (c) sized carbon fibres using scanning electron microscope (SEM) images taken in SEI mode. There are noticeably large differences in the means shown in the fifth band of the table, at each stage of processing. However, the standard deviations overlap in all cases indicating that there is no notable statistical evidence for any difference at any stage of processing. The properties are compared for these fibres in figure 6.7(d), specifically the atomic ratio (O/C) and surface energy versus ILSS for unoxidised, oxidised, and epoxy-sized carbon fibres. It is clear from this figure that there is no correlation between properties and roughness for these fibres, as the sized fibres are lower in roughness yet exhibit the highest ILSS values. This indicates that other forms of adhesion can play a more significant role in contributing to interfacial strength and that mechanical interlocking should never simply be 'assumed' without further investigation into the chemical compatibility and wettability of the fibre (see chapter 3).

In the final band of the table, R_a roughness values are shown for unsized Magnamite® IM6 fibres (0%) and Magnamite® IM6 fibres surface treated by an electrolytic anodisation process to 20%, 100%, 200%, and 600% of their nominal sizing level. There is an apparent increase in roughness on fibre surfaces when anodisation is above 100%, and this in turn is a result of surface etching from the anodic process [16]. While the examples from each band in the list from table 6.3 are not exhaustive, they represent a range of differences commonly observed in reinforcing fibres.

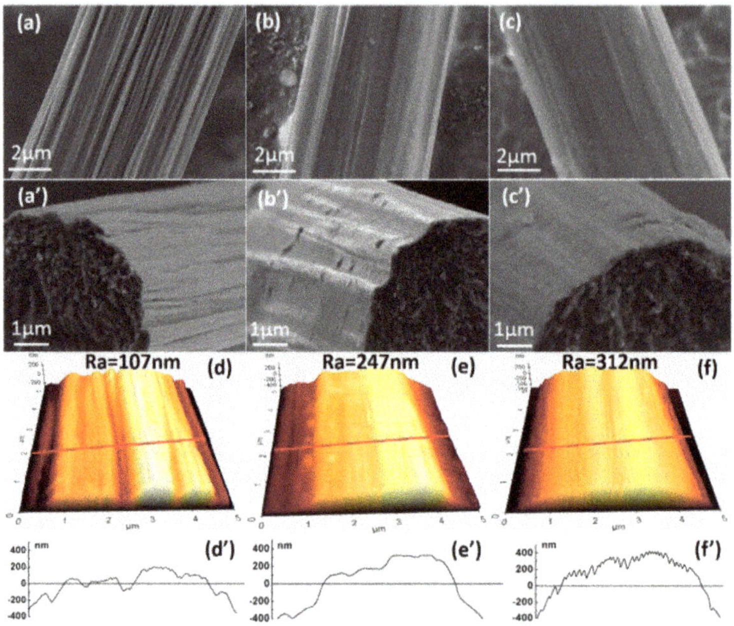

Figure 6.8. SEM images of surface morphologies of carbon fibres (CFs): desized CF (a and a'), E-1-sized CF (b and b'), and F-1-sized CF (c and c'); AFM images and corresponding height profiles along the red line of CF, desized CF (d and d'), E-1-sized CF (e and e'), and F-1-sized (f and f'). Reprinted from [17]. Copyright (2018), with permission from Elsevier.

A clean visual example of how sizing can alter surface morphology and roughness on a carbon fibre surface is shown in figure 6.8 [17], where desized carbon fibre is shown (a and a'), E-1-sized carbon fibre is shown (b and b'), and F-1-sized carbon fibre is shown (c and c'). AFM images and their corresponding height profiles along the red line of carbon fibre are shown in d and d' for desized carbon fibre, in e and e' for E-1-sized carbon fibre, and in f and f' for F-1-sized carbon fibre. Here, E-51 refers to bisphenol A epoxy resin 618 sizing and F-51 refers to bisphenol F epoxy resin NPEF-164X (Dalian liansheng trading Co., Ltd, China).

References

[1] ISO 21 920-1:2021 Geometrical product specifications (GPS)—Surface texture: Profile Part 1: Indication of surface texture, International Organization for Standardization, Geneva, Switzerland

[2] ISO 21 920-2:2021 Geometrical product specifications (GPS)—Surface texture: Profile Part 2: Terms, definitions and surface texture parameters, International Organization for Standardization, Geneva, Switzerland

[3] ISO 21 920-3:2021 Geometrical product specifications (GPS)—Surface texture: Profile Part 3: Specification operators, International Organization for Standardization, Geneva, Switzerland

[4] ISO 8785:1998 Geometrical Product Specification (GPS)—Surface imperfections—Terms, definitions and parameters, International Organization for Standardization, Geneva, Switzerland

[5] ISO 16 610-21:2011 Geometrical product specifications (GPS)—Filtration Part 21: Linear profile filters: Gaussian filters, International Organization for Standardization, Geneva, Switzerland

[6] Gao X, Gillespie J W Jr, Jensen R E, Li W, Haque B Z and McKnight S H 2015 Effect of fiber surface texture on the mechanical properties of glass fiber reinforced epoxy composite *Composites A* **74** 10–7

[7] Gao S L and Mader E 2002 Characterisation of interphase nanoscale property variations in glass fibre reinforced polypropylene and epoxy resin composites *Composites A* **33** 559–76

[8] Zhuang R C, Burghardt T, Plonka R, Liu J W and Mader E 2010 Affecting glass fibre surfaces and composite properties by two stage sizing application *eXPRESS Polym. Lett.* **4** 798–808

[9] Dey M, Deitzel J M, Gillespie J W Jr and Schweiger S 2014 Influence of sizing formulations on glass epoxy interphase properties *Composites A* **63** 59–67

[10] Song W, Gu A, Liang G and Yuan L 2011 Effect of the surface roughness on interfacial properties of carbon fibers reinforced epoxy resin composites *Appl. Surf. Sci.* **257** 4069–74

[11] Meng L H, Chen Z W, Song X L, Liang Y X, Huang Y D and Jiang Z X 2009 Influence of high temperature and pressure ammonia solution treatment on interfacial behavior of carbon fiber/epoxy resin composites *J. Appl. Polym. Sci.* **113** 3436–41

[12] Ruan R, Cao W and Xu L 2020 Quantitative characterization of physical structure on carbon fiber surface based on image technique *Mater. Des.* **185** 108225

[13] Gao S L, Mader E and Zhandarov S F 2004 Carbon fibers and composites with epoxy resins: topography, fractography and interphases *Carbon* **42** 515–29

[14] Dilsiz N and Wightman J P 1999 Surface analysis of unsized and sized carbon fibers *Carbon* **37** 1105–14

[15] Kafi A, Huson M, Creighton C, Khoo J, Mazzola L, Gengenback T, Jones F and Fox B 2014 Effect of surface functionality of PAN-based carbon fibres on the mechanical performance of carbon/epoxy composites *Compos. Sci. Technol.* **94** 89–95

[16] Drzal L T, Sugiura N and Hook D 1997 The role of chemical bonding and surface topography in adhesion between carbon fibers and epoxy matrices *Compos. Interfaces* **4** 337–54

[17] Liu F, Shi Z and Dong Y 2018 Improved wettability and interfacial adhesion in carbon fibre/epoxy composites via an aqueous epoxy sizing agent *Composites A* **112** 337–45

IOP Publishing

Composite Interfaces in Mechanical Design

Parvez Alam

Chapter 7

Mechanical testing—methods and standards

7.1 Overview of the chapter

This chapter aims to detail different interface test methods for composites at different length scales. The chapter will begin by discussing some of the macroscale methods including lap shear tests, double cantilever beam (DCB) tests, tapered double cantilever beam (TDCB) tests, wedge-peel and impact wedge-peel tests, end-notch flexure (ENF) tests, end-loaded split tests, four-point end-notched flexure tests, edge ring crack torsion (ERCT) tests, and mixed mode flexure (MMF) tests. Following this, the second part of this chapter will consider fibre and particulate scale tests including single fibre pullout tests, micro-bond tests, micro-indentation tests, Broutman tests, and fibre fragmentation tests.

7.2 Macro-scale test methods

7.2.1 Lap shear test

Lap shear tests are commonly used to judge the shear strength of joined composites. Example standards for this test where fibre-reinforced plastics (FRPs) are at least one of the adherends include ASTM D5868 [1], Standard test method for lap shear adhesion for FRP bonding, and ISO 22841:2021 and its amendment ISO 22841:2021/Amd 1:2022 [2, 3], Composites and reinforcements fibres: Carbon fibre-reinforced plastics and metal assemblies—Determination of the tensile lap shear strength. The test is primarily designed to assess the shear strength of adhesively bonded composites, and as such, the test method favours joints rather than co-cured composite parts. When designing lap shear joints, it is important to try to design the joint optimally, such that it is stronger than the materials that are being joined, as this will ensure the materials are used to their maximum mechanical capacity. In the simplest single shear lap joint, figure 7.1, the adherends are sufficiently rigid and we can assume idealised uniform movement in the adhesive, which in turn results in uniform shear stress across the length of the adhesive. The

doi:10.1088/978-0-7503-5688-6ch7

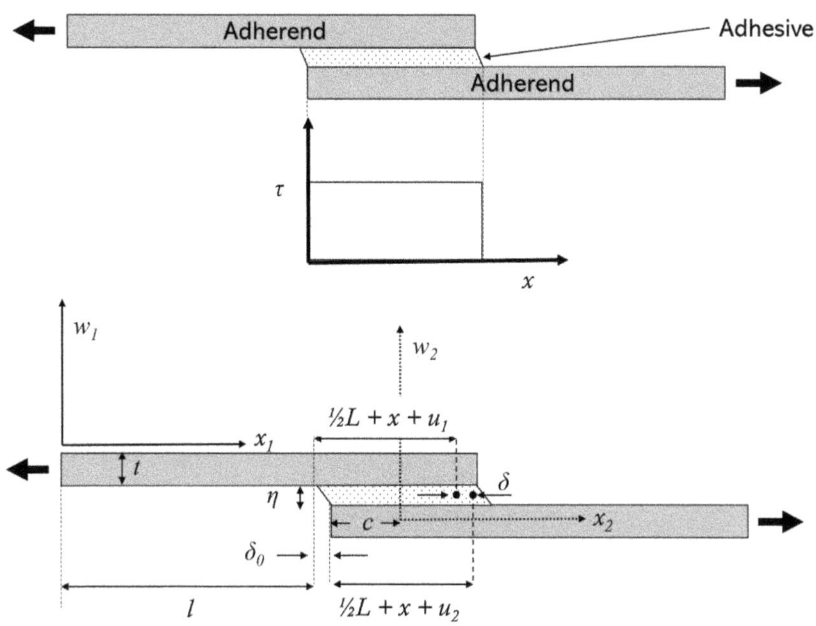

Figure 7.1. Fundamental scheme of an ideal single shear lap specimen showing an equally distributed shear stress across the entire adhesive (top/middle) and a deformed single shear lap specimen (bottom) that considers possible non-uniform (variable) shear stresses on deformation (bottom).

shear stress, τ, between adherend and adhesive is simply calculated as $\tau = G\gamma$, where G is the shear modulus of the adhesive and γ is the shear strain under deformation calculated as $\gamma = \frac{\delta}{\eta}$, where δ is the deformational change on load in the adhesive and η is the thickness of the adhesive layer.

In composites, many lap shear joints are rigid in the adherends. Nevertheless, there can be cases where adherends are not rigid and this can give rise to non-uniform stresses across the interface, resulting in peeling of adherend from adhesive. In a case such as this, the variability in stress across the interface can be accounted for using Volkersen's shear lag model [4], broken down in equation (7.1) (see bottom image in figure 7.1 for reference), where u_1 and u_2 are the changes in deformation in x of each of the adherends and are calculated as $u_1 = \int_{-\frac{1}{2}L}^{x} \varepsilon_1 dx$ and $u_2 = \int_{-\frac{1}{2}L}^{x} \varepsilon_2 dx$; ε is normal strain in the adherend; L is lap length; and x is the distance from the centre of the adhesive where shear stress is calculated, [5].

$$\tau = G\gamma = G\frac{\delta}{\eta} = G\frac{\delta_0 + u_1 - u_2}{\eta} \tag{7.1}$$

When a lap shear joint is tensioned, as in figure 7.1, the bending moment, M, contributes significantly to the stress magnitude arising from the diagonal loading path from adherend to adherend via the adhesive. Figure 7.2 summarises the free body loads experienced in a single lap shear joint. Here, V is shear loading and T is

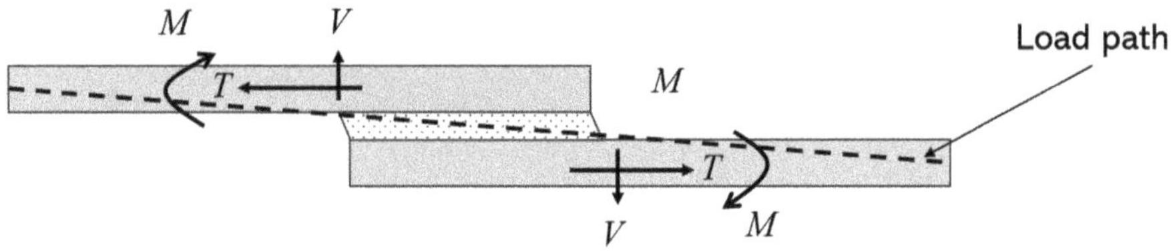

Figure 7.2. Single lap shear joint loading shown as a free body diagram: V is shear, M is the moment, and T is tension.

tensile loading. As a result of the mixed modes of loading, adhesives can peel and larger bending moments may magnify the extent of adhesive-adherend peeling.

The Goland–Reissner [6] analysis on the rotation of adherends in single lap shear joints is the earliest analytical model on this phenomenon and is presented in equation (7.2). Here, M_0 is the moment per unit width (at the ends of the overlap); P is the applied tension; t is the adherend thickness; c is the length of the overlap from the centre of the overlap; and ξ is the square root of the ratio between load and the unit-width bending stiffness of the adherend D such that $x_i = \sqrt{\frac{P}{D}}$. D can be computed as $D = \left(\frac{E}{1-\nu}\right)\frac{t^3}{12}$, where E is the elastic modulus of the adherend and ν is its Poisson's ratio. The boundary conditions for the Goland–Reissner model are as follows: (i) $w_1(x_1 = 0) = 0$ (ii) $w_1(x_1 = 1) = w_2(x_2 = -c)$, (iii) $\frac{dw_1(x_1 = l)}{dx_1} = \frac{dw_2(x_2 = -c)}{dx_2}$, and (iv) $w_2(x_2 = 0) = 0$, [7]. While the Goland–Reissner analysis is the basis for several modified computations for lap shear, it does not account for the adhesive stiffness, which will have its own effect on the mechanics of a lap shear joint. The Hart–Smith [8] model improves the mechanical analysis on a single lap shear by taking adhesive thickness, η, into account. The Hart–Smith model is given in equation (7.3), where l_a is the length of the free adherend from the end of the joint overlap to the loading point. The boundary conditions for the Hart–Smith model are as follows: (i) $w_1(x_1 = l) = w_2(x_2 = -c)$, (ii) $\frac{dw_1(x_1 = l)}{dx_1} = \frac{dw_2(x_2 = -c)}{dx_2}$, (iii) $\frac{d^2w_1(x_1 = l)}{dx_1^2} = \frac{d^2w_2(x_2 = -c)}{dx_2^2}$, and (iv) $\frac{d^4w_1(x_1 = l)}{dx_1^4} = \frac{d^4w_2(x_2 = -c)}{dx_2^4} = \frac{t+\eta}{2D}\frac{d\tau(x_2 = -c)}{dx_2}$ [7].

$$M_0 = \frac{Pt}{2}\frac{1}{1 + 2\sqrt{2}\tanh\left(\frac{\xi c}{2\sqrt{2}}\right)} \tag{7.2}$$

$$M_0 = \frac{P(t+\eta)}{2}\frac{1 + \frac{\eta}{t}}{1 + \frac{\xi c}{\tanh(\xi l_a)} + \frac{\xi^2 c^2}{6}} \tag{7.3}$$

M_0 is generally expressed according to equation (7.4), where k is the bending moment factor. The Goland–Reissner bending moment factor k_{GR} is expressed by equation (7.5) and the Hart–Smith bending moment factor k_{HS} by equation (7.6).

$$M_0 = \frac{Pt}{2}k \tag{7.4}$$

$$k_{\mathrm{GR}} = \frac{1}{1 + 2\sqrt{2}\tanh\left(\frac{\xi c}{2\sqrt{2}}\right)} \tag{7.5}$$

$$k_{\mathrm{HS}} = \frac{1 + \frac{\eta}{t}}{1 + \frac{\xi c}{\tanh(\xi l_a)} + \frac{\xi^2 c^2}{6}} \tag{7.6}$$

7.2.2 Mode I—Double cantilever beam (DCB) test

The DCB test is widely used amongst composites engineers. With an analysis based on linear elastic fracture mechanics (LEFM), it is a simple test to execute, providing quantitative measurements of mode I fracture initiation and propagation under either static or cyclic loading. A common test setup is shown in figure 7.3 (from [9]). To make the DCB coupon, one thick section composite coupon is fitted with a Teflon insert at the mid-section of one of the coupon ends prior to curing. The Teflon insert prevents curing at the site of the insert and a linear pseudo-crack is made within the DCB coupon, which is then attached at the crack-end to metallic hinges that are used to apply load to the coupon and open the crack in mode I. ASTM D5528-13 [10] is an often used standard for FRP specimens and was the first

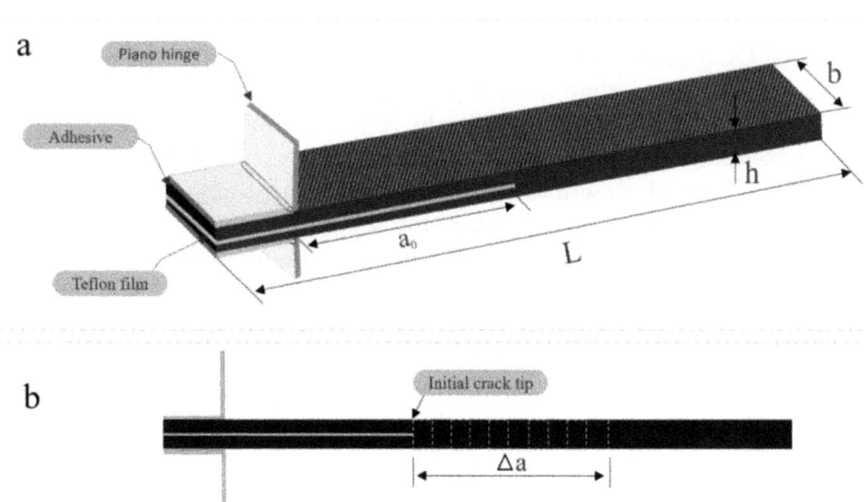

Figure 7.3. Schematic representation of the DCB specimen: (a) with piano hinges; (b) initial crack tip. Reproduced from [9]. CC BY 4.0.

standardisation body to describe DCB test sample dimensions. Another common standard in use is ISO 15 024:2023 [11].

Crack opening occurs when the strain energy release rate, G_I, reaches a critical value for strain energy release, G_{IC}, and this is fundamentally determined using the energy method described by equation (7.7), where U is the strain energy, b is the specimen width, and a is the crack length.

$$G_I = G_{\mathrm{IC}} = -\frac{1}{b}\frac{dU}{da} \tag{7.7}$$

In addition to the energy method (see equation (7.7)), compliance methods can be used, in particular the compliance calibration equation (7.8) or modified compliance calibration equation (7.9), where P is the load; δ is the load point displacement; C is the compliance of the DCB specimen calculated as $\frac{\delta}{P}$; n is the slope of $\log C$ against $\log a$; h is the specimen thickness; and A_1 is the slope from a plot of $\frac{a}{b}$ against $C^{\frac{2}{3}}$. Finally, the modified beam method can also be simplified to determine the crack opening critical strain energy release, G_{IC}, in accordance with equation (7.10) [25].

$$G_{\mathrm{IC}} = \frac{nP\delta}{2ba} \tag{7.8}$$

$$G_{\mathrm{IC}} = \frac{3P^2C^{\frac{2}{3}}}{2A_1bh} \tag{7.9}$$

$$G_{\mathrm{IC}} = \frac{3P\delta}{2ba} \tag{7.10}$$

7.2.3 Mode I—Tapered double cantilever beam (TDCB) test

The Tapered DCB (TDCB) method, like the DCB method, is a LEFM method for analysing the fracture initiation and propagation in materials. While there are currently no TDCB standards specifically geared towards the design of fibre-reinforced composites, a good generic TDCB standard worth visiting is the ISO 25 217:2009 [12], which describes the method for structurally adhered joints. The TDCB test is shown in figure 7.4. Here, the setup is similar to the DCB test (see figure 7.3); however, the composite material tapers out at an angle to a wider end. The TDCB method produces more accurate results than the DCB method, as tapering the composite from the region of the crack decreases the stress concentration at the crack tip [13]. The result of the TDCB method is a constant rate of change of compliance with crack length, over a large range of crack lengths, and to ensure the representation of this, a geometry factor, m, is used, equation (7.11).

$$m = \frac{3a^2}{h^3} + \frac{1}{h} \tag{7.11}$$

Three methods are used when determining the value of G_{IC} in TDCB samples. These include those predicted by (i) simple beam theory, (ii) the experimental

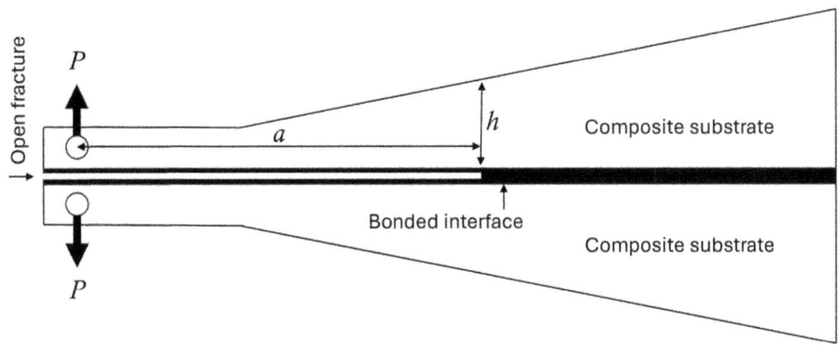

Figure 7.4. Schematic representation of a TDCB specimen.

compliance method, and (iii) the corrected beam theory. The strain energy release of a TDCB specimen is represented by equation (7.12), and when adhesive layers are thin in accordance with **simple beam theory,** $\frac{dC}{da} = \frac{8}{Eb}(\frac{3a^2}{h^3} + \frac{1}{h})$, where E can be either the Young's modulus or the flexional modulus of the composite material. As such, equation (7.11) together with the substitution for $\frac{dC}{da}$ can be plugged into equation (7.12) to yield equation (7.13).

$$G_{\mathrm{IC}} = \frac{P^2}{2b}\frac{dC}{da} \tag{7.12}$$

$$G_{\mathrm{IC}} = \frac{4P^2}{Eb^2}\left(\frac{3a^2}{h^3} + \frac{1}{h}\right) = \frac{4P^2}{Eb^2}m \tag{7.13}$$

When using the **experimental compliance method** a correction is advised when load blocks are used as it includes the displacement correction, F, and a load-block correction, N, expressed as a ratio to yield the strain energy release shown in equation (7.14). Here, if piano hinges are used or drilled holes, then $N = 1$ and F only has importance when the ratio $\frac{\delta}{a}$ is large (typically >0.4).

$$G_{\mathrm{IC}} = \frac{nP\delta}{2ba}\frac{F}{N} \tag{7.14}$$

Corrected beam theory is applied when compliance from deflection and rotation occurs at the beam root, which corresponds to the built-in crack tip, equation (7.15). These effects are more important than those that are typically associated with shear deformations when both compliance and G_{IC} are predicted from TDCB specimens, such as those manufactured using metal substrates [14].

$$G_{\mathrm{IC}} = \frac{4P^2m}{Eb^2}\left[1 + 0.43\left(\frac{3}{ma}\right)^{\frac{1}{3}}\right] \tag{7.15}$$

7.2.4 Mode I—Wedge-peel (WP) and impact wedge-peel (IWP) tests

The WP test is still not commonplace amongst composites engineers. The method is fundamentally a method for assessing adhesion between substrates, and as such, use of the method in composites engineering tends to be limited to adhered composites. This said, authors such as Sasheeth *et al* [17] do show that the WP test can be used for ply-to-ply peel testing at constant velocity and note that the WP test results in a higher overall force than other peel test methods, such as T-peel tests. The IWP test is essentially the same in terms of its physical setup as the WP test, except that instead of using a constant loading velocity, the wedge is impacted to generate cracking between adhered substrates. Figure 7.5 from [16] shows a schematic illustration of an IWP specimen in (a) an IWP specimen in a drop tower chamber in (b) and an example of an impact from a striker to the wedge shackle shoulder leading to cleavage of a composite IWP coupon in (c).

In a WP test, the WP strength S_{WP} is typically calculated in accordance with equation (7.16), where P_{max} is the imposed maximum loading and b is the specimen width.

$$S_{WP} = \frac{P_{max}}{b} \tag{7.16}$$

In an IWP, the relevant standard ISO 11 343:2019 [18] recommends an impaction rate of at least 2 $m\ s^{-1}$ and an impact energy of at least 50 J, using a drop tower setup, a servo-hydraulic test machine, or a pendulum impaction rig. This standard identifies three main outputs of interest from the test. The first is a cleavage force, which is the momentary force recorded during stable crack growth, the second is the average cleavage force, which is an average measure of force between the first 25% and last 10% of the IWP curve, and the third is the dynamic cleavage energy, which is the energy required to cause failure between two adhered substrates. While the

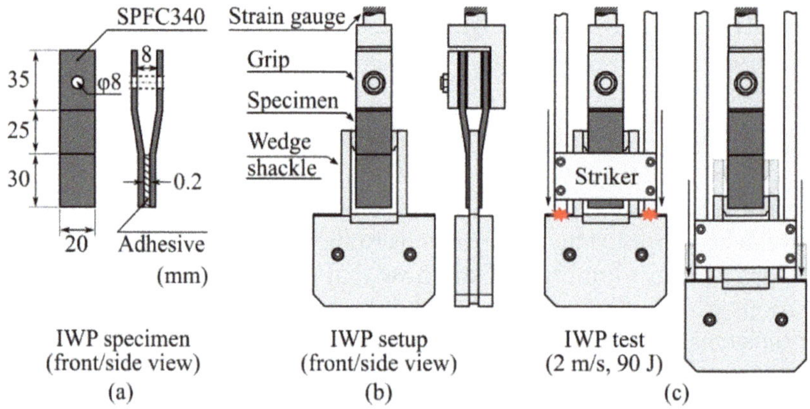

Figure 7.5. Schematic illustration of (a) an IWP specimen, (b) an IWP setup in the drop tower chamber, and (c) example of an initial impact of the striker to the wedge shackle shoulder (left) and impact cleavage process (right) during an IWP test. Reproduced from [16]. CC BY 4.0.

standard considers the case of stable crack growth, Blackman *et al* [19] note that there are cases of unstable crack growth in some IWP tests, which occurs when there is no plateau region that ISO 11 343 considers in its IWP analysis. In such cases, ISO 11 343 therefore fails to provide an accurate measurement of impact behaviour and alternative schemes are needed, such as recognising short-time scale (<7 ms [19]) failures as unstable, and the conversion of these values to null values.

7.2.5 Mode II—End notched flexure (ENF) test

The end notched flexure (ENF) test is a standarised test method commonly used to determine the mode II delamination toughness of composite laminates. A useful standardised ENF method for unidirectional continuous FRPs can be found in the ASTM D7905/D7905M-19e1 [15]. The test setup involves the 3-point loading of an end-notched coupon, which is essentially structurally similar to a DCB coupon. The method is shown in the schematic of figure 7.6, where a_0 is the initial distance from an outer roller located under the crack tip to the crack tip, S is the distance between the two outermost rollers, the specimen length is L, and h and b are the specimen thickness and width, respectively. Loading causes mode II shearing about the crack, resulting in a stress concentration at the crack tip, and mode II shear failure from the crack tip along the length of the coupon. This occurs when the strain energy release rate G_{II} reaches a critical value for the release of strain energy, G_{IIc}, a newer derivation of which is covered by De Moura [20] based on a **compliance-based beam method (CBBM)**, equations (7.17)–(7.23). In equation (7.17), the strain energy of a specimen due to bending is shown where the effects of shear are included. Here, E_f is the flexural modulus of the composite along its primary axis, M_f is the bending moment, τ is the shear stress, and G_{13} is the shear modulus in the $_{13}$ plane, while b and h are the specimen width and thickness, respectively.

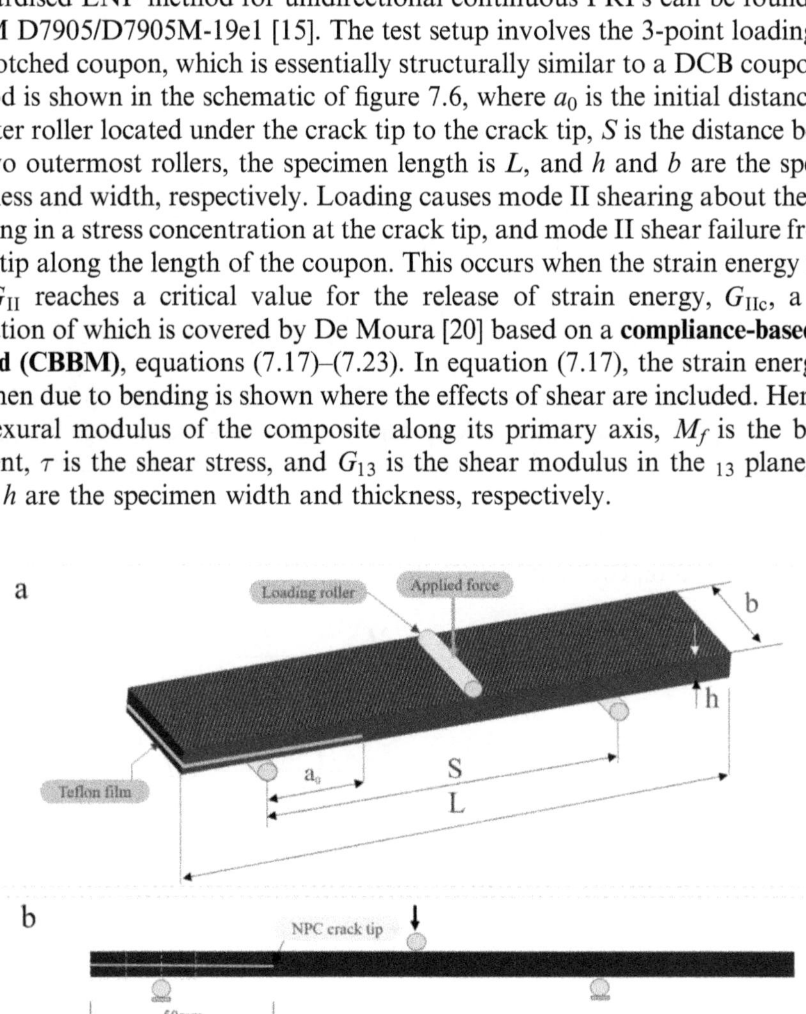

Figure 7.6. End notched flexure (ENF) specimen configuration: (a) specimen nomenclature; (b) crack tip location. Reproduced from [9]. CC BY 4.0.

$$U = \int_0^{2L} \frac{M_f^2}{2E_f I} dx + \int_0^{2L} \int_{-h}^{h} \frac{\tau^2}{2G_{13}} bdydx \tag{7.17}$$

In equation (7.17), the shear stress is expressed according to equation (7.18), where A_i is the area of cross-section of the beam, $c_i = 0.5\,\text{h}$, and represents the thickness of the beam, and V_i is the transverse load of the i segment ($0 \leqslant x \leqslant a$, $a \leqslant x \leqslant L$, or $L \leqslant x \leqslant 2L$).

$$\tau = \frac{3}{2} \frac{V_i}{A_i} \left(1 - \frac{y^2}{c_i^2}\right) \tag{7.18}$$

The displacement δ at the point of loading for the crack length a can be derived from Castigliano's theorem and is shown in equation (7.19).

$$\delta = \frac{dU}{dP} = \frac{P(3a^3 + 2L^3)}{8E_f bh^3} + \frac{3PL}{10G_{13}bh} \tag{7.19}$$

E_f is a fundamental material property that critically relates load, P, and displacement, δ, and it can be calculated using the parameters of initial crack length, a_0, and initial compliance, C_0, equation (7.20).

$$E_f = \frac{3a_0^3 + 2L^3}{8bh^3} \left(C_0 - \frac{3L}{10G_{13}bh}\right)^{-1} \tag{7.20}$$

De Moura [20] further stipulates the need for a crack tip correction based on the fact that the region ahead of the crack tip experiences micro-cracks, fibre bridging, an inelastic deformations. This region ahead of the crack tip is termed the fracture process zone (FPZ) and the real crack length, a_{eq}, is considered in view of compliance, C, equation (7.21), where a_{FPZ} is the crack length based on the FPZ, and consequently a_{eq} is calculated based on equation (7.22), where $C_{\text{corr}} = C - \frac{3L}{10G_{13}bh}$.

$$C = \frac{3(a + \Delta a_{\text{FPZ}})^3 + 2L^3}{8E_f bh^3} + \frac{3L}{10G_{13}bh} \tag{7.21}$$

$$a_{\text{eq}} = a + \Delta a_{\text{FPZ}} = \left[\frac{C_{\text{corr}}}{C_{0\text{corr}}} a_0^3 + \frac{2}{3}\left(\frac{C_{\text{corr}}}{C_{0\text{corr}}} - 1\right) L^3\right] \tag{7.22}$$

Leading to a corrected calculation of G_{IIc} in equation (7.23).

$$G_{\text{II}} = G_{\text{IIc}} = \frac{9a_{\text{eq}}^2 P^2}{16E_f b^2 h^3} \tag{7.23}$$

The CBBM method discussed by De Moura is a simple and useful method by which means G_{IIc} can be determined in an ENF test. Some older methods are still used, and these include the **compliance calibration method (CCM)** and beam theory

(BT). CCM is one of the more oft-used models for calculating G_{IIc}, equation (7.24), using values for load, P, displacement, δ, and crack length, a, and the constant m, which has a cubic relationship with the compliance C such that $C = D + ma^3$, where D is also a constant.

$$G_{\text{IIc}} = \frac{3P^2\, ma^2}{2b} \tag{7.24}$$

Beam theory methods are also oft-used to determine G_{IIc} subjected to ENF testing and are described in ASTM D7905/D7905M-19e1 [21], which recommends the use of both **direct beam theory methods (DBTM)**, equation (7.25), and **corrected beam theory methods (CBTM)**, equation (7.26), with the axial composite modulus calculated as $E_f = \frac{L^3}{4bh^3 C_0}$, where C_0 is the initial compliance at $a = a_0$.

$$G_{\text{IIc}} = \frac{9a^2 P\delta}{2b(2L^3 + 3a^3)} \tag{7.25}$$

$$G_{\text{IIc}} = \frac{9a^2 P^2}{16b^2 h^3 E_f} \tag{7.26}$$

7.2.6 Mode II—End-loaded split (ELS) test

The end-loaded split (ELS) test uses a similar coupon structure to that of a DCB specimen (also similar in construct to an ENF coupon). The coupon is end clamped such that the end clamp itself rests on free-sliding bearings and can thus translate axially. Unlike the DCB test, an ELS test applies load directly over the crack (or delamination) via a single end block, causing the structure to bend, figure 7.7. The P–δ relationship follows similar steps to that of the ENF specimens as described by De Moura [20] and is shown in equation (7.27).

$$\delta = \frac{dU}{dP} = \frac{P(3a^3 + L^3)}{2bh^3 E_1} + \frac{3PL}{5bhG_{13}} \tag{7.27}$$

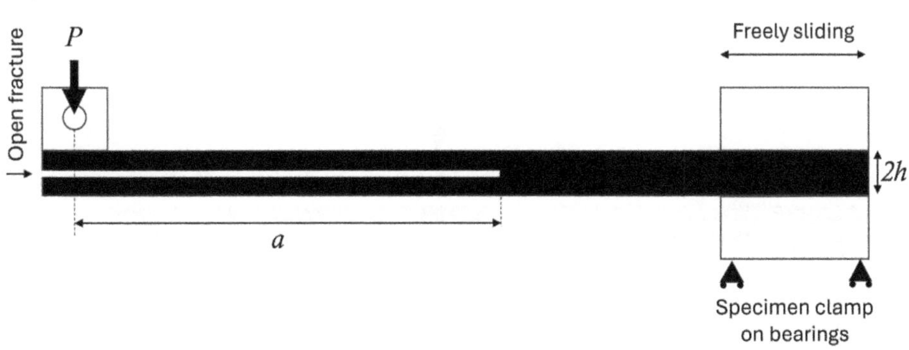

Figure 7.7. Schematic of an end-loaded split (ELS) test configuration.

The initial compliance C_0, which can be measured experimentally, is given by equation (7.28), where L_{ef} is an effective beam length that should consider rotational effects at the root region of clamping, alongside details of crack tip stresses/strains not ordinarily included in beam theory. The influence of the FPZ on the real crack length Δa_{FPZ} during crack propagation is considered in the compliance, C, as shown in equation (7.29). By combining equations (7.28) and (7.29), the equivalent crack length a_{eq} can be deduced (equation (7.30)) and G_{IIc} can be obtained, equation (7.31), which can be obtained without measurement of the crack as it propagates, since this equation relies on only load and displacement during crack propagation.

$$C_0 - \frac{3a_0^3}{2bh^3E_1} = \frac{L_{ef}^3}{2bh^3E_1} + \frac{3L_{ef}}{5bhG_{13}} \tag{7.28}$$

$$C - \frac{3(a + \Delta a_{FPZ})^3}{2bh^3E_1} = \frac{L_{ef}^3}{2bh^3E_1} + \frac{3L_{ef}}{5bhG_{13}} \tag{7.29}$$

$$a_{eq} = a + \Delta a_{FPZ} = \left[(C - C_0)\frac{2bh^3E_1}{3} + a_0^3\right]^{\frac{1}{3}} \tag{7.30}$$

$$G_{IIc} = \frac{9P^2a_{eq}^2}{4b^2h^3E_1} \tag{7.31}$$

7.2.7 Mode II—Four-point end-notched flexure (4ENF) test

The 4ENF test is a relatively recent test method introduced by Martin and Davidson [22], a scheme for which is shown in figure 7.8. The coupons are prepared in a similar fashion to those for DCB and ELS tests, albeit without any attached load blocks. 4ENF samples are subjected to four-point bending loads and the G_{IIc} is calculated according to equation (7.32), where $C = C_0a + C_1$, as is specific for 4ENF samples [23].

$$G_{IIc} = \frac{P^2}{2b}\frac{dC}{da} \tag{7.32}$$

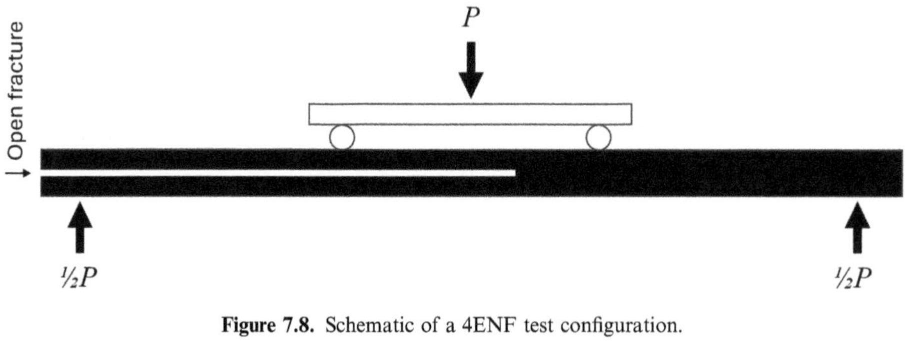

Figure 7.8. Schematic of a 4ENF test configuration.

7.2.8 Mode III—Edge ring crack torsion (ERCT) test

Experimentally, it is extremely difficult to achieve pure mode III delamination in composites. It is for this reason that there is significantly less in the literature on mode III testing, compared to modes I and II testing. A common mistake is to assume that mode III delamination is analagous to shear delamination since through testing, it can be shown that $G_{IIc} \neq G_{IIIc}$ [25]. Optimally therefore, a mode III test should minimise any mode I and mode II deformations. For this reason, the numerous tests proposed since the 1980s have not really gained much popularity, since none of these have proven to eliminate mode II behaviour within a mode III test.

One of the more successful mode III tests is the ERCT test, which develops a pure mode III fracture, while concurrently only developing small variations of G_{III} along the crack front [24]. The test setup is shown in figure 7.9, where a flat composite plate is placed between two torsional devices, which localise rotation to the z-axis of the plate. A crack ring is further introduced around the edge of the specimen, to additionally localise the fracture in a specific plane in the z-axis. G_{III} can be calculated in accordance with equation (7.33), where K_{III} is a semi-analytical stress intensity factor calculated from equation (7.34), where $f(\frac{d}{D})$ is determined from equation (7.35); T is the axial torque; D is the diameter including an edge ring crack at a cross section under T; and d is the crack front circular diameter [26].

$$G_{III} = \frac{1}{2G} K_{III}^2 \tag{7.33}$$

$$K_{III} = \frac{16T}{\pi d^3} \sqrt{\pi \frac{D-d}{2}} \cdot f\left(\frac{d}{D}\right) \tag{7.34}$$

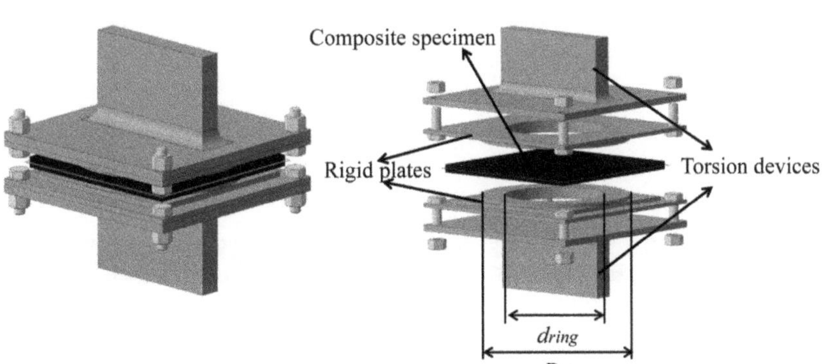

Figure 7.9. Example of an ERCT test. Reprinted from [24]. Copyright (2016), with permission from Elsevier.

$$f\left(\frac{d}{D}\right) = \frac{3}{8}\sqrt{\frac{d}{D}} \cdot \left[1 + \frac{1}{2}\frac{d}{D} + \frac{3}{8}\left(\frac{d}{D}\right)^2 + \frac{5}{16}\left(\frac{d}{D}\right)^3 + \frac{35}{128}\left(\frac{d}{D}\right)^4 + 0.208\left(\frac{d}{D}\right)^5\right] \quad (7.35)$$

7.2.9 Mixed mode (modes I and II) flexure (MMF)/bending (MMB)

A relevant standard for MMF tests, also referred to as mixed mode bending (MMB), is the ASTM D6671/D6671M-19 [27], which has been developed for unidirectional continuous FRPs. One version of the test scheme is shown in figure 7.10. This setup allows for the simultaneous application of a tensile pull and a 3-point bend to the crack. Variations of the setup may include the use of piano hinges or tensile loading blocks (see [25]), but the essential criterion is that both tension and bending are applied to the induced crack.

Mixed mode failure (a combination of crack opening and crack shearing—modes I and II, respectively) occurs when the strain energy release rate ($G_{I/II}$) from both modes reaches a critical value, $G_{(I/II)c}$ (equation (7.36)), which is the sum of both the G_{Ic} (equation (7.37)) and G_{IIc} (equation (7.38)), where c is the lever length of the MMF/MMB test apparatus, χ is a crack correction parameter, and E_{1f} is the modulus of elasticity in the fibre direction measured in flexure and is shown in equation (7.39). In equation (7.39), m is the slope of the load–displacement curve, a_0 is the initial crack length, and C_{sys} is the system compliance.

$$G_{(\mathrm{I/II})\mathrm{c}} = G_{\mathrm{Ic}} + G_{\mathrm{IIc}} \quad (7.36)$$

$$G_{\mathrm{Ic}} = \frac{12P^2(3c - L)^2}{16b^2h^3L^2E_{1f}}(a + \chi h)^2 \quad (7.37)$$

$$G_{\mathrm{IIc}} = \frac{9P^2(c + L)^2}{16b^2h^3L^2E_{1f}}(a + 0.42\chi h)^2 \quad (7.38)$$

$$E_{1f} = \frac{8(a_0 + \chi h)^3(3c - L)^2 + [6(a_0 - 0.42h\chi)^3 + 4L^3](c + L)^2}{16L^2bh^3\left(\frac{1}{m} - C_{sys}\right)} \quad (7.39)$$

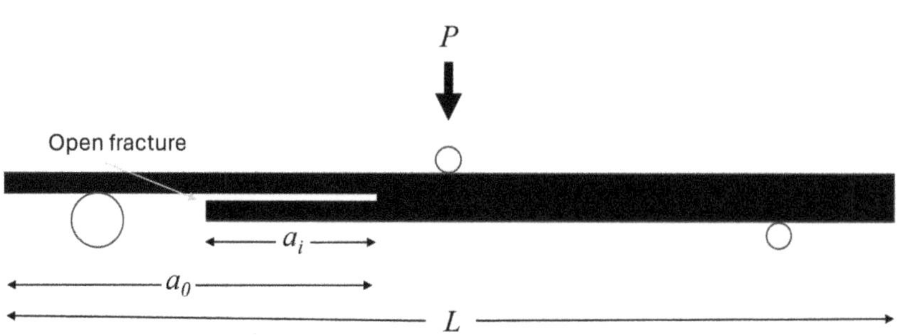

Figure 7.10. Scheme of a MMF/MMB test.

7.3 Fibre and particulate scale test methods

7.3.1 Single fibre pullout

Single fibre pullout is an oft-used test method for the determination of fibre–matrix bond strength. The test setup was initially developed by Penn [30] using the Broutman test method [28] as a basis for design. A generic test setup is shown in figure 7.11, where prior to testing, a fibre is embedded into a block of matrix material and attached to a fibre mount using an appropriately stiff and strong adhesive to ensure force transfer from the mount to the fibre. The gauge length of the setup is the distance between the edge of the mount (assuming the fibre is bonded all

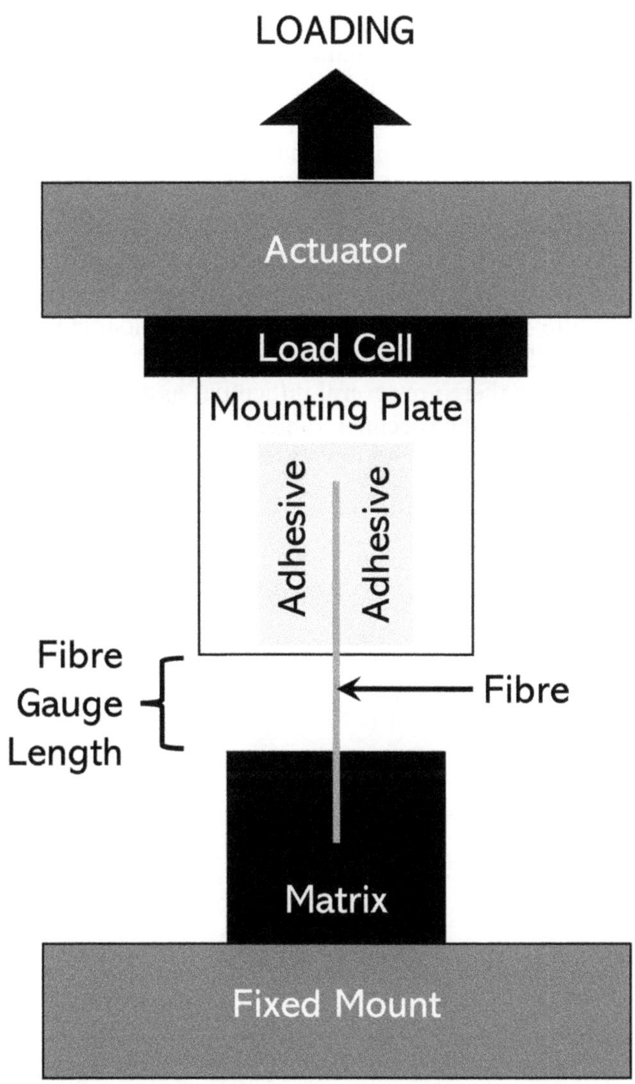

Figure 7.11. Typical experimental setup for a single fibre pullout test.

the way to the mount-edge) and the surface of the substrate. Load is applied through an actuator and the force measured over pulling time as the fibre pulls out of the matrix material.

On loading, the fibre is pulled in tension and as such, the fibre may stretch or fracture in tension if it is too deeply embedded into the matrix, or, the matrix material may deform excessively if it is insufficient in terms of its either radial and/or vertical dimensions. Ideally, a pullout test would assume that two rigid bodies interact solely at the interface during loading, and test parameters therefore seek to minimise (or account for) bulk material deformation during a test. As such, there are critical parameters that determine the success of a fibre pullout test. The first is the embedding length of the fibre into the matrix (a limiting length), which should be sufficient such that the fibre is completely pulled out from the matrix (i.e. it is not so long that the fibre itself ends up stretching or fracturing). The second is that the dimensions of the matrix block should be large enough to prevent large deformations of the block in any direction. Appropriate models are based on the understanding of mode II (shear) failure, and the base equation for bond strength in shear, τ_{bond}, takes into account the shear surface such that $\tau_{\text{bond}} = \frac{P_{\text{fl}}}{\pi dl}$, where P_{fl} is the fracture load of the fibre from the matrix, d is the diameter of the fibre, and l is the length of the embedded part of the fibre [25]. Friction-based shear sliding of the fibre from the matrix occurs at the tail end of the load–displacement curve, figure 7.12. While mode II failure is calculated, the initial mode of loading is mode I, due to Poisson contractions of the fibre from the matrix.

At small length scales there are several difficulties associated with the tests. This includes high variability in the specimen dimensions due to fibre size distributions,

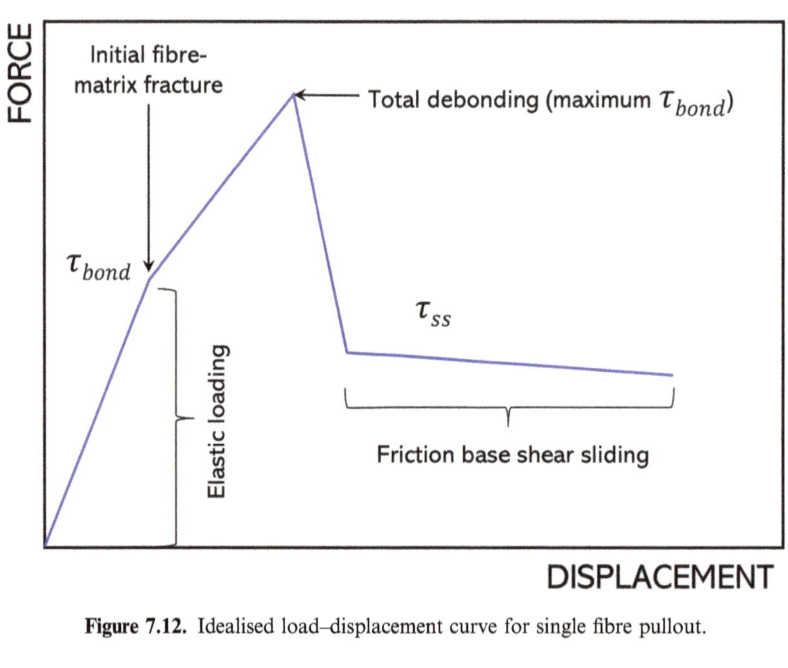

Figure 7.12. Idealised load–displacement curve for single fibre pullout.

which can be significant in natural fibres. The wetting characteristics of the fibre surface can additionally create menisci which, depending on size, can create complications by decreasing stress uniformity. Due to the nature of the materials used and their small sizes, ambient thermal fluctuations can introduce large variations in the radial stresses of the fibre in the matrix. Finally, these tests typically experience large levels of scatter, presumably as a result of the prior mentioned difficulties.

7.3.2 Micro-bond

The micro-bond test is a specialised form of the single fibre pullout test. Here, a single fibre is pulled out from a droplet of resin that is cured axisymmetrically on the fibre. A pair of knife edges are loaded parallel to the fibre axis to shear debond the droplet from the fibre, figure 7.13. The shear bond strength, τ_{bond}, is essentially calculated in the same way as for the single fibre pullout test, and is based on mode II failure criteria, such that $\tau_{\mathrm{bond}} = \frac{P_{fl}}{\pi d l}$, where P_{fl} is the fracture load of the fibre from the matrix, d is the diameter of the fibre, and l is the length of the embedded part of the fibre (the droplet length). It should nevertheless be noted that the initial loading of the fibre in the matrix is mode I, which is fundamentally due to Poisson contraction of the fibre from the matrix material.

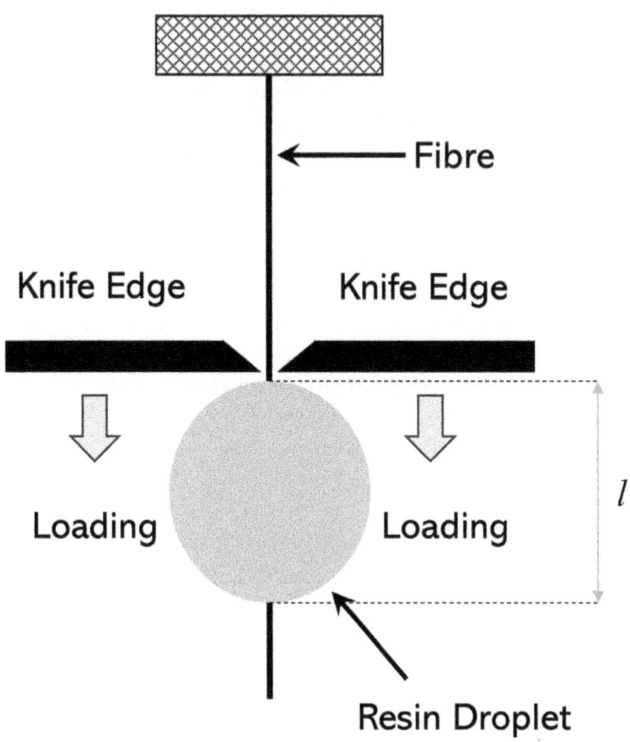

Figure 7.13. Typical experimental setup for a single fibre micro-bond test.

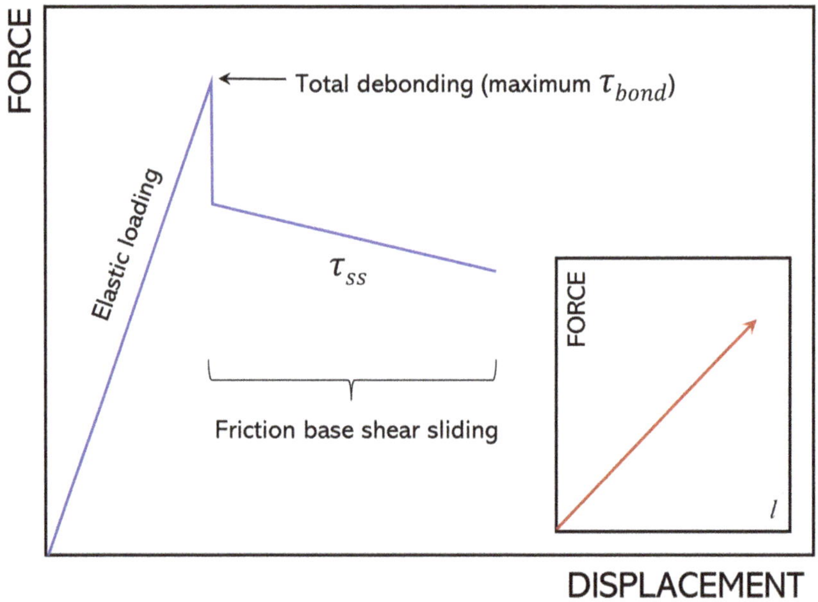

Figure 7.14. Idealised load–displacement curve for single fibre micro-bond test (main graph) and force against the droplet length, l (inserted graph).

The force–displacement curve of a micro-bond test is often linear elastic to failure, with shear sliding possible if the resin remains intact and the loading rate is slow enough to enable the monitoring of sliding, figure 7.14 (main graph). τ_{bond} also generally increases as a function of increasing l, figure 7.14 (insert). The results from micro-bond tests can vary considerably because the test is sensitive to variations in the test setup, such as knife pair spacing, shape, sharpness and rigidity of the loading knives, and the size and shape of the droplet and its meniscus [31, 32]. Variability in the spacing distances between the knife pairs, for example, gives rise to different stress concentrations [33], which result from the position of knife contact on the surface of the droplet. This is important as meniscus failure is typically deemed to invalidate the test, and the chances of meniscus failure can be increased through the position of knife pair loading on the droplet and shape of the knives, relative to the shape and size of the droplet.

7.3.3 Micro-indentation

The micro-indentation method, originally proposed by [34], is a method that involves pressing either a flat-ended or a rounded-end indentor onto the end of a fibre embedded into matrix material. The force is a compression loading onto the surface of the fibre, which translates to shearing of the fibre–matrix interface, leading to fibre debonding followed by slip. Fibre expansion does take place due to Poisson's effects, and as such, the fibre–matrix interface shearing is not a pure mode II deformation as radial expansion of the fibre under loading also induces radial compression into the matrix. To ensure indentation is as perfectly transverse to the

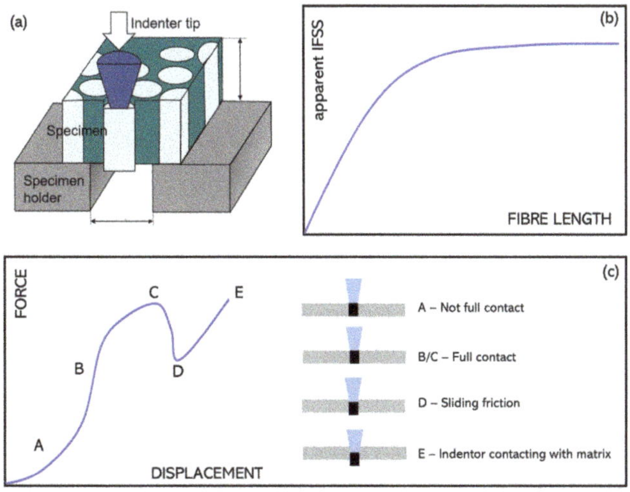

Figure 7.15. (a) Typical experimental setup for an indentation fibre-push through test ((a) reprinted from [35], CC BY 4.0), (b) expected profile for force plotted against increasing fibre length, and (c) the different stages of fibre indentation shown against a force–displacement curve.

fibre axis as possible, the surface is often polished flat prior to indentation. While fibre debonding can be inferred from the force–displacement curve, acoustic emission events can also be traced to identify the debonding. A typical experimental setup (from [35]) is shown in figure 7.15(a).

The apparent interfacial shear stress (apparent IFSS) is calculated as $\mathrm{IFSS} = \frac{F_{max}}{2\pi rl}$, where F_{max} is the maximum indentation force, r is the fibre radius, and l is the embedded fibre length. Fracture toughness is essentially a combination of compression (Poisson effects) and shear and is calculated as $K_{I/IIc} = \frac{E_S}{2\pi rl}$, where E_s is the total separation energy of the fibre from the matrix. Generally speaking, the apparent IFSS increases rapidly with fibre length, tapering off to a plateau as the fibre stress transfer aspect ratio is reached, figure 7.15(b) [36]. Additionally, there are different stages in an indentation curve, which relate to the different stages of indentation, as shown in figure 7.15(c) [37]. Initially there is no or incomplete contact (stage A). When full contact is reached, the resistance to loading increases and there is an elastic portion of the curve that can taper off to a maximum (stages B–C, where C is the maximum load and the point of debonding). Sliding friction of the fibre within the matrix follows the debonding stage (D), and as the indentor presses into the matrix, there is an increase in loading (E) and the test should be stopped by this stage.

A point of note is that the indentor size and shape will affect loading, and thus the non-uniformity of the stress state at the interface. In addition, stress state non-uniformity will also be affected by the position of the indentor tip on the fibre face, which would ideally be at the planar centroid of the fibre face; however, there are practical challenges to achieving this. The size and shape of the indentor can affect the extent to which damage occurs within the fibre through contact with the indentor. Damage to the fibre through indentation can affect the reliability of the output data from indentation tests, and is one cause of statistical scatter within sample sets.

7.3.4 Broutman test

The Broutman test [28] is another test designed to determine the fibre–matrix interface strength. This test is different than more common tests such as micro-debond and single fibre pullout, as loading is applied to the matrix in the Broutman test rather than to the fibre. The fundamental test setup is shown in figure 7.16(a), where a single fibre is embedded into matrix material that tapers inward from the ends to the centre (essentially, a curved neck specimen). Broutman's logic in developing a curved neck specimen was to encourage a tensile debond between fibre and matrix (due to Poisson effects) rather than a debond through shearing. The compressive loading on the matrix material results in a Poisson's ratio-induced transverse expansion. This expansion is typically greater than any expansion of the individually embed reinforcing fibre. Consequently, interfacial tension, *S*, is generated to conserve the continuum structure and the interfacial tension at debond can be represented according to equation (7.40), where σ_m is the axial stress on the minimum section; ν_m and ν_f are the Poisson ratios of matrix and fibre, respectively;

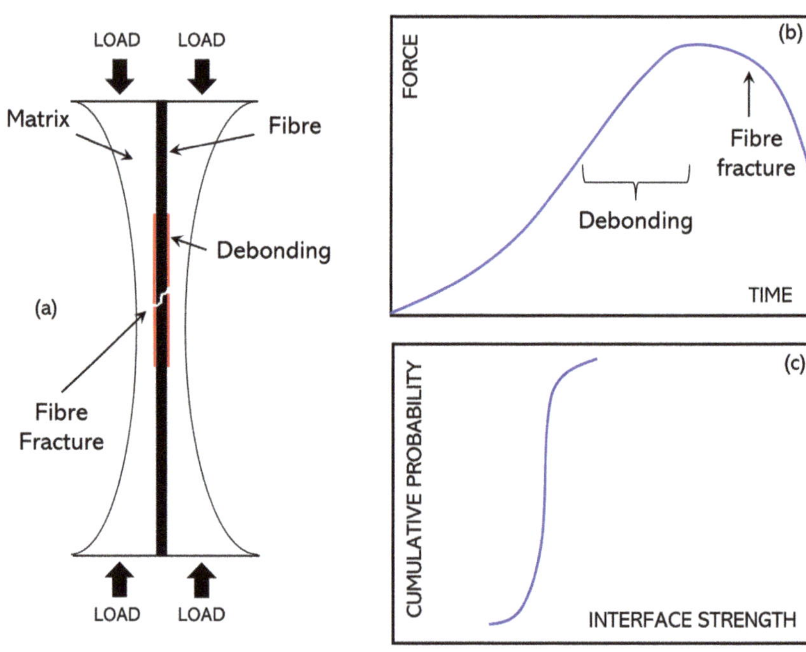

Figure 7.16. (a) Typical experimental setup for a Broutman test, (b) representative force–time curve for a Broutman test indicating general regions of debonding and initial fibre fracture, and (c) a representative cumulative probability–interface strength curve for a Broutman test.

and E_m and E_f are the elastic moduli of matrix and fibre materials, respectively. A representative load–time curve for a Broutman test is shown in figure 7.16(b) where the regions of debonding and initial fibre fracture are indicated, and a representative curve of cumulative probability against interface strength is shown in figure 7.16(c).

$$S = -\frac{\sigma_m(\nu_m - \nu_f)E_f}{(1 - nu_m)E_f + 1(1 - \nu_f - 2\nu_f^2)E_m} \tag{7.40}$$

7.3.5 Fibre fragmentation test

Another fibre–matrix test where loading is applied from matrix to fibre, rather than the more common fibre to matrix, is the fragmentation test. In this test, a single fibre is embedded into a matrix, which is tensioned until the fibre fragments; it is a comprehensive test procedure that is available through the Risoe National Laboratory, Denmark [29]. The scheme for this test is shown in figure 7.17(a). In figure 7.17(b), a typical crack density versus applied strain curve is shown, and figure 7.17(c) illustrates a typical set of curves showing the number of fragments plotted against strain. In this test, when the shear bond strength between fibre and matrix is high, the number of fibre fractures (and hence fragments) is higher and these fibre fragments are oftentimes also shorter. When the shear bond strength is

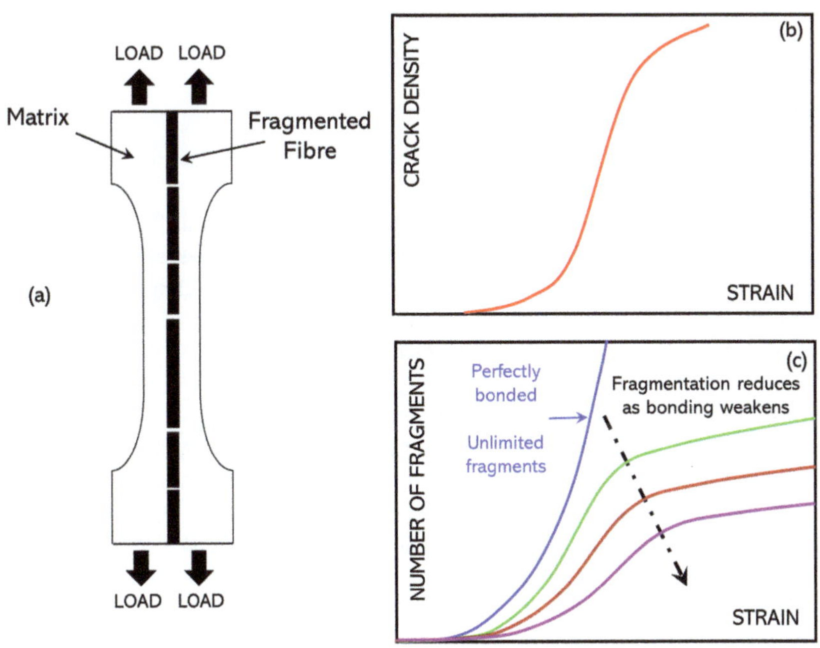

Figure 7.17. (a) Typical experimental setup for a fibre fragmentation test, (b) representative crack density versus applied strain curve for a fibre fragmentation test, and (c) a representative number of fragments versus strain curve for a fibre fragmentation test. Here, it can be noted that fibre fragmentation decreases as fibre–matrix bonding weakens.

reduced, fibre–matrix debonding occurs, sparing the fibre, reducing the number of fibre fragments, and thus creating larger fibre fragments [38]. In a theoretically perfectly bonded situation, fibre fragments can be theoretically infinite in number and infinitely small.

Following the Risoe procedure [29], the average interfacial shear strength, τ_{if}, can be calculated for bonded, debonded, or matrix yielded situations according to equation (7.41), where σ_f is the fibre strength at a critical length, l_c, and d is the diameter of the fibre. Here, l_c is calculated according to equation (7.42), where the average fibre length, $\bar{l}$, is measured through testing as an average fibre fragmentation length at a point where there are no further breaks in the fibre with increasing strain. Additionally, the fibre strength at the fragment length can be determined using a Weibull distribution, equation (7.43), where m is the Weibull modulus and σ_0 is the characteristic fibre strength at L_0, the gauge length. The characteristic fibre strength σ_0 at a new gauge length, L_1, i.e. ($\sigma_0(L_1)$), is then determined using equation (7.43).

$$\tau_{if} = \frac{\sigma_f(l_c)d}{2l_c} \tag{7.41}$$

$$l_c = \frac{4}{3}\bar{l} \tag{7.42}$$

$$\sigma_0(L_1) = \sigma_0(L_0)\left(\frac{L_0}{L_1}\right)^{\frac{1}{m}} \tag{7.43}$$

References

[1] ASTM D5868-01 2023 Standard Test Method for Lap Shear Adhesion for Fiber Reinforced Plastic (FRP) Bonding

[2] ISO 22 841 2021 Composites and reinforcements fibres—Carbon fibre reinforced plastics (CFRPs) and metal assemblies—Determination of the tensile lap-shear strength

[3] ISO 22 841:2021/Amd 1 2022 Composites and reinforcements fibres—Carbon fibre reinforced plastics(CFRPs) and metal assemblies—Determination of the tensile lap-shear strength—Amendment 1: Precision data General information

[4] Volkersen O 1938 Die nietkraftverteilung in zugbeanspruchten nietverbindungen mit konstanten laschenquerschnitten *Luftfahrtforschung* **15** 41–7

[5] Crocombe A D and Ashcroft I A 2008 Simple lap joint geometry *Modeling of Adhesively Bonded Joints* ed L F Martins da Silva and A Ochsner (Berlin: Springer)

[6] Goland M and Reissner E 1944 The stresses in cemented joints *J. Appl. Mech.* **11** A17–27

[7] Redmann A, Damodaran V, Tischer F, Prabhakar P and Osswald T A 2021 Evaluation of single-lap and block shear test methods in adhesively bonded composite joints *J. Compos. Sci.* **5** 27

[8] Hart-Smith L J 1973 Adhesive bonded single lap joints *Contractor Report* (Washington, DC: NASA)

[9] Ali A, Andriyana A, Hassan S B A and Ang B C 2021 Fabrication and thermo-electro and mechanical properties evaluation of helical multiwall carbon nanotube-carbon fiber/epoxy composite laminates *Polymers* **13** 1437

[10] ASTM D5528/D5528M-21 2022 *Standard Test Method for Mode I Interlaminar Fracture Toughness of Unidirectional Fiber-Reinforced Polymer Matrix Composites* (West Conshohocken, PA: ASTM International) 2013

[11] ISO 15 024:2023 Fibre-reinforced plastic composites—Determination of mode I interlaminar fracture toughness, GIC, for unidirectionally reinforced materials, International Organization for Standardization, Geneva, Switzerland

[12] ISO 25 217:2009 Adhesives—Determination of the mode 1 adhesive fracture energy of structural adhesive joints using double cantilever beam and tapered double cantilever beam specimens, International Organization for Standardization, Geneva, Switzerland

[13] Tsang W L 2020 The use of tapered double cantilever beam (TDCB) in investigating fracture properties of particles modified epoxy *SN Appl. Sci.* **2** 751

[14] Blackman B R K, Hadavinia H, Kinloch A J, Paraschi M and Williams J G 2003 The calculation of adhesive fracture energies in mode I: revisiting the tapered double cantilever beam (TDCB) test *Eng. Fract. Mech.* **70** 233–48

[15] ASTM D7905/D7905M-19e1 2019 *Standard Test Method for Determination of the Mode II Interlaminar Fracture Toughness of Unidirectional Fiber-Reinforced Polymer Matrix Composites.* (West Conshohocken, PA: ASTM International)

[16] Baek D, Sim K-B and Kim H-J 2021 Mechanical characterization of core–shell rubber/epoxy polymers for automotive structural adhesives as a function of operating temperature *Polymers* **13** 734

[17] Satheesh B, Tonejc M, Potakowskyj L *et al* 2018 Peel strength characterisation on ply/ply interface using wedge and T-peel/pull-type tests *Polymers Polym. Compos.* **26** 431–45

[18] ISO 11 343:2019 Adhesives—Determination of dynamic resistance to cleavage of high-strength adhesive bonds under impact wedge conditions—Wedge impact method, International Organization for Standardization, Geneva, Switzerland

[19] Blackman B R K, Kinloch A J, Taylor A C and Wang Y 2000 The impact wedge-peel performance of structural adhesives *J. Mater. Sci.* **35** 1867–84

[20] De Moura M F S 2008 Interlaminar mode II fracture characterization *Delamination Behaviour of Composites, Woodhead Publishing Series in Composites Science and Engineering* ed S Sridharan (Sawston: Woodhead Publishing (Imprint of Elsevier)) pp 310–26

[21] ASTM D7905/D7905M-19e1 *Standard Test Method for Determination of the Mode II Interlaminar Fracture Toughness of Unidirectional Fiber-Reinforced Polymer Matrix Composites* (West Conshohocken, PA: ASTM International) 2019

[22] Martin R H and Davidson B D 1999 Mode II fracture toughness evaluation usingfour point bend, end notched flexure test *Plast. Rubber Compos.* **28** 401–6

[23] Wang W X, Nakata M, Takao Y and Matsubara T 2009 Experimental investigation on test methods for mode II interlaminar fracture testing of carbon fiber reinforced composites *Composites A* **40** 1447–55

[24] Ge Y, Gong X, Hurez A and De Luycker E 2016 Test methods for measuring pure mode III delamination toughness of composite *Polym. Test.* **55** 261–8

[25] Alam P 2021 *Composites Engineering: An A-Z Guide* (Bristol: IOP Publishing)

[26] Tada H, Paris P C and Irwin G R 1973 *The Stress Analysis of Cracks Handbook* (Del Research Corporation) 34 p

[27] ASTM D6671/D6671M-19 2019 *Standard Test Method for Mixed Mode I-Mode II Interlaminar Fracture Toughness of Unidirectional Fiber Reinforced Polymer Matrix Composites* (West Conshohocken, PA: ASTM International)

[28] Broutman L J 1969 Measurement of the fiber-polymer matrix interfacial strength *Interfaces in Composites* (Philadelphia, PA: American Society for Testing and Materials (ASTM)) 27–41 pp
[29] Feih S, Wonsyld K, Minzari D, Westermann P and Lilholt H 2004 Testing procedure for the single fiber fragmentation test Risoe National Laboratory. Denmark. Forskningscenter Risoe. Risoe-R No. 1483(EN)
[30] Penn L S 1981 A new approach to surface energy characterization for adhesive performance prediction *Surf. Interface Anal.* **3** 161–4
[31] Nishikawa M, Okabe T, Hemmi K and Takeda N 2008 Micromechanical modeling of the microbond test to quantify the interfacial properties of fiber-reinforced composites *Int. J. Solids Struct.* **45** 4098–113
[32] Wada A and Fukuda H 1999 Microbond test for the fibre/matrix interfacial shearing strength *Proceedings of the 1999 Int. Conf. on Composite Materials (Paris, France, 5–9 July 1999)*
[33] Mendels D A, Leterrier Y and Manson J A E 2001 The influence of internal stresses on the microbond test-I: theoretical analysis *J. Compos. Mater.* **36** 257–384
[34] Mandell J F, Chen J H and McGarry F J 1980 A microdebonding test for in-situ fiber–matrix bond and moisture effects *Research Report* R80-1 (Department of Materials Science and Engineering, Massachusetts Institute of Technology)
[35] Rohrmuller B, Gumbsch P and Hohe J 2021 Calibrating a fiber–matrix interface failure model to single fiber push-out tests and numerical simulations *Composites A* **150** 106607
[36] Al-Ostaz A, Drzal L T and Schalek R L 2004 Fibre-matrix adhesion measured by the micro-indentation test: data reduction based on real time simulation *J. ASTM Int.* **1** JAI11934
[37] Godara A, Gorbatikh L, Kalinka G, Warrier A, Rochez O, Mezzo L, Luizi F, van Vuure A W, Lomov S V and Verpoest I 2010 Interfacial shear strength of a glass fiber/epoxy bonding in composites modified with carbon nanotubes *Compos. Sci. Technol.* **70** 1346–52
[38] McCarthy E D and Soutis C 2019 Determination of interfacial shear strength in continuous fibre composites by multi-fibre fragmentation: a review *Composites A* **118** 281–92

IOP Publishing

Composite Interfaces in Mechanical Design

Parvez Alam

Chapter 8

DMA of composite interfaces

8.1 Introduction

Unlike with the various reinforcement/matrix interface tests discussed in chapter 7, there is comparatively very little in terms of standard practice for the determination of dynamic adhesive properties between reinforcement and matrix. Yet, given that strain rate-dependent materials are often used in the manufacture of composites, there is benefit in discussing the methods that have been used to determine the adhesive strength of reinforcement to matrix under dynamic loading conditions. A method that has been used successfully is dynamic mechanical analysis (DMA). DMA is a commonly used dynamic polymer characterisation method used in polymer science as it relates molecular structure, processing, and properties to polymer behaviour, figure 8.1. Nevertheless, it has gained popularity amongst composites engineers as a method for interface characterisation, since polymer molecules connected at interfaces, and their molecular arrangements within interphases formed through the presence of reinforcing, are often very different than bulk polymer materials, exhibiting different properties and behaviours. These new properties and behaviours are detectable through DMA, and this chapter aims, therefore, to build both a basic understanding of the principles underlying DMA methods, as well as to introduce the utility of this method in interface characterisation.

8.2 DMA: the fundamentals

8.2.1 General concepts

The DMA method involves the use of a DMA instrument, which imposes mechanical loads to a material and induces small deformations. The responses the material has to the imposed deformation is typically a function of time in a DMA, but temperature is also used in many DMA instruments to assess the response as a function of temperature. In such cases, the analysis is a dynamic mechanical thermal

doi:10.1088/978-0-7503-5688-6ch8

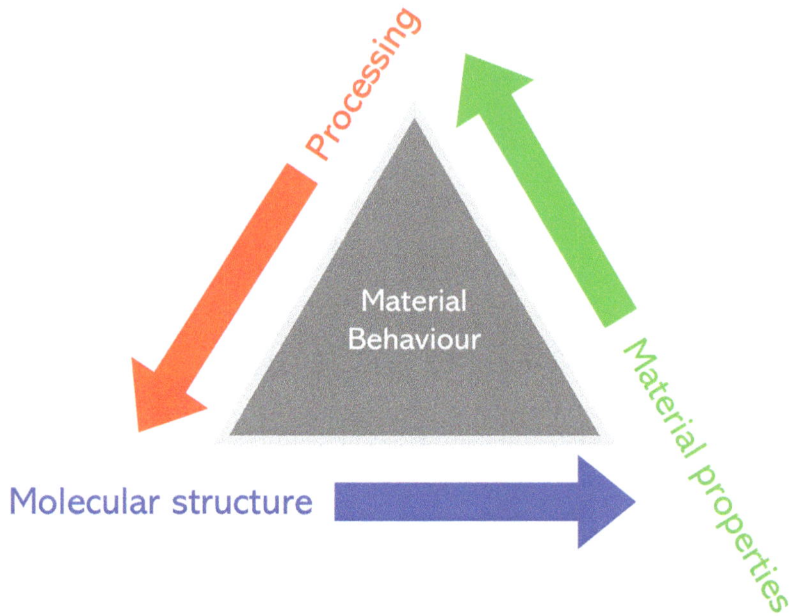

Figure 8.1. The molecular structure of a material, the conditions for its processing/manufacture, and its properties can all be related to material behaviour using DMA techniques.

analysis. While static loading is possible in a DMA, the more common mode of loading is oscillatory as it enables the easy determination of viscoelastic properties in the material. A viscoelastic material has the inherent properties of both an elastic solid and a viscous liquid. An elastic solid follows Hooke's law, equation (8.1), where E is the elastic modulus, σ is stress, and ε is strain; while a viscous fluid follows Newton's law of viscosity, equation (8.2), where η is the fluid viscosity and $\frac{d\varepsilon}{dt}$ is the strain rate. Viscoelastic behaviour is a combination of both the Hookean solid and the Newtonian fluid model, and is shown in equation (8.3). This behaviour can be captured by DMA through strain or stress applied to the material. When an oscillatory stress or strain (as these are proportional at small deformations) is applied to a material, it produces a response over time, t. These can be illustrated as deformation and response curves in figure 8.2, where we also note that the phase angle, δ, is used to describe a shift as a function of time between the deformation curve and the response curve. In a purely Hookean material, $\delta = 0°$, while in a purely viscous fluid, $\delta = 90°$. Since a viscoelastic material exhibits both Hookean and viscous characteristics, it can thus exhibit a wider range of phase angles such that $0° < \delta < 90°$. The extent to which a material behaves like a pure Hookean solid (i.e. the in-phase component) is represented as $\sigma' = \sigma^*\cos(\delta)$, while the extent to which a material behaves like an ideal liquid (i.e. the out-of-phase component) is represented as $\sigma'' = \sigma^*\sin(\delta)$, where σ^* is referred to as the complex stress, equation (8.4).

$$E = \frac{\sigma}{\varepsilon} \tag{8.1}$$

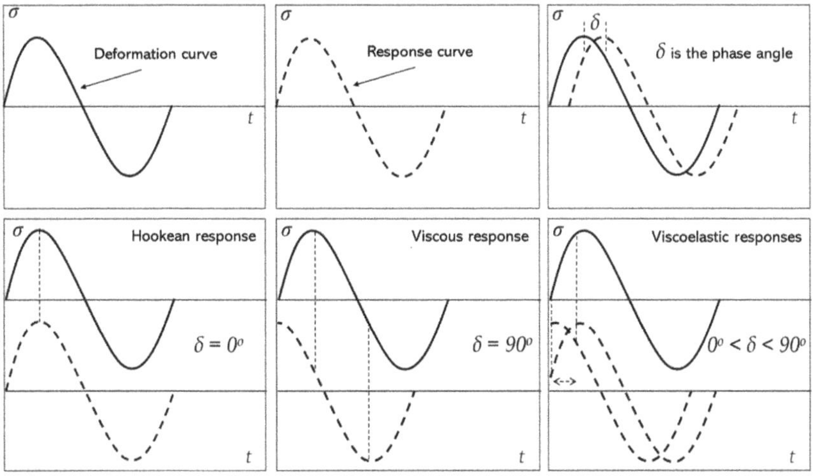

Figure 8.2. The deformation and response curves in a σ–t context showing both resultant deformation and response curves. The shift seen when these curves are out of phase with respect to time is the phase angle, δ, which is $\delta = 0°$ for pure Hookean solids, $\delta = 90°$ for pure viscous fluids, and $0° < \delta < 90°$ for viscoelastic materials.

$$\sigma = \eta \frac{d\varepsilon}{dt} \tag{8.2}$$

$$\sigma = E\varepsilon + \eta \frac{d\varepsilon}{dt} \tag{8.3}$$

$$\sigma^* = \sigma' + i\sigma'' \tag{8.4}$$

8.2.2 Properties of viscoelastic materials

From the relationships described, certain important parameters can be calculated, such as the complex modulus, E^*, which is the resistance of the material to deformation, equation (8.5); the storage modulus, E', which is measure of the energy storage capacity of the material, equation (8.6); the loss modulus, E'', which informs on the material ability to dissipate (or lose) energy, equation (8.7); and tan δ, also referred to as the 'loss factor', which is a measure of material damping and is a ratio between the loss and storage moduli, equation (8.8).

$$E^* = \frac{\sigma^*}{\varepsilon} \tag{8.5}$$

$$E' = \frac{\sigma^*}{\varepsilon} \cos \delta \tag{8.6}$$

$$E'' = \frac{\sigma^*}{\varepsilon} \sin \delta \tag{8.7}$$

$$\tan\delta = \frac{E''}{E'} \tag{8.8}$$

8.2.3 Material behaviour in a temperature sweep

Both frequency sweeps (progressive ramping of the loading cycles per unit time) and temperature sweeps (progressively ramping the temperature per unit time) are used to analyse E', E'', and $\tan\delta$. Temperature sweeps can be particularly useful as the E', E'', and $\tan\delta$ curves from temperature sweeps can be used to extract detailed molecular-level information about the material. Figure 8.3 shows a typical outcome on E', E'', and $\tan\delta$ curves in a temperature sweep. The initial low temperature range represents the glassy region of the curves where E' is typically high and E'' is typically low. This glassy region indicates that the molecules are locked (almost 'frozen') in place, and as the temperature increases, E' begins to decrease indicating the unlocking of molecules within the material. As the temperature enters a glass transition region, there is a steep decline in E', which indicates the large-scale unlocking or unfreezing of molecules, i.e. increase molecular mobility. This glass transition region is essentially a reversible state between a harder glassy material state and a softer amorphous (gel like/rubbery) material state. It is within the glass transition region that we can extract information on the glass transition temperature, Tg. The glass transition temperature is often taken as the peak of the $\tan\delta$ curve, but in fact this curve indicates a range of different extant Tg values within the material, the peak itself indicating the predominant Tg. As such, the breadth of the $\tan\delta$ curve indicates how heterogeneous a material is. A material exhibiting only one Tg will appear as a single peak (i.e. with 'no' width), a straight line at the relevant Tg, and any increase in the width of the curve from this indicates that there is greater heterogeneity. The width is most often taken at the half-peak height. The area under the $\tan\delta$ curve and the height are indicators of the amount of energy a

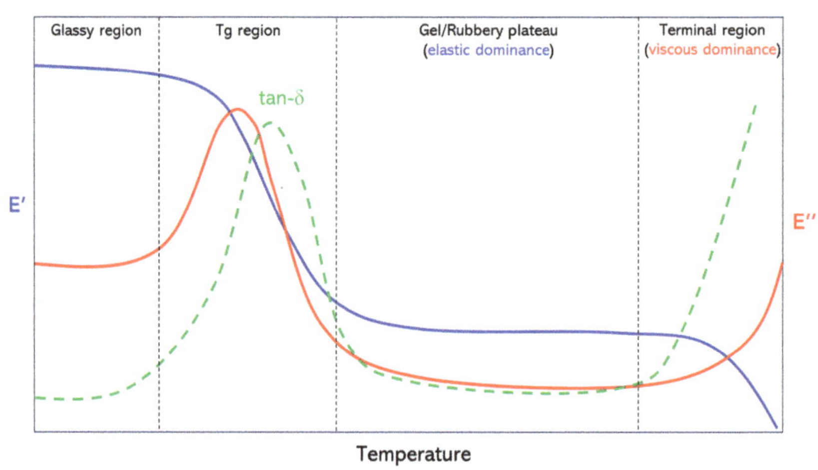

Figure 8.3. Typical E', E'', and $\tan\delta$ curves during a temperature sweep, indicating the different regions resulting from progressive increases in temperature.

material can absorb, and hence the effectiveness of material damping. As the area under the tan δ curve increases, so too does the level of molecular mobility, and thus damping properties (higher absorption and dissipation properties). An increase in the value of tan δ indicates an increase in the dissipation properties of a material, and as such, the lower the tan δ value, the more elastically it acts. The gel/rubbery region indicates how amorphous the material is. Over the rubbery plateau seen in figure 8.3, the material can be strained significantly (several hundred percent strain), and can return back to its original length on release, i.e. it shows elastic dominance where large elastic deformation is possible under low stresses. This region is below the melt region and is where molecular-level entanglements and crosslinking can occur. The storage modulus E' in the rubbery region can be used as a qualitative indicator of entanglement and crosslinking density, with higher values of E' indicating higher levels of molecular entanglement and crosslinking. The final region is the terminal region (above the melt temperature) where the material acts essentially as a viscous fluid, and there is, relatively speaking, little to no elastic response.

8.3 Characterisation of interfacial interactions using DMA

DMA can be used as a characterisation tool when there are sufficiently strong interfacial interactions between matrix and reinforcement. Since DMA is strongly influenced by the conformations and interactions at the molecular level, it is able to detect the effects that reinforcements have on molecules at interfaces where molecular pinning can occur [1, 2], and can also detect the subsequent knock-on effects that pinned interface molecules have on interphase molecular structures, arrangements, and interactions. The strength of the interactions will depend on a number of factors [3], some of which have been considered in table 8.1.

8.3.1 Observation of E' and E'' shifts

A good starting point in determining interface effects in composites using DMA methods is to observe shifts in E' and E''. When considering this as an option, it is important to ensure that the only independent variable is the interface itself. Since the DMA method is very sensitive to the effects of fibre orientation [4], reinforcement volume fraction [5, 6], processing conditions [7, 8], and environmental influences [9], care must be taken to minimise variation in aspects besides those introduced by modifications to the interfaces. Fundamentally, the effects of the interfaces on both storage and loss moduli can be most effectively compared assuming all material components of the composite, manufacturing constraints, environmental test conditions, and reinforcement orientations are the same. Variations to, for example, reinforcement treatments, which affect molecular pinning at interfaces as well as interphase development, will then be observable in the DMA outputs for E' and E''.

Figure 8.4 shows by illustration idealised curves that can be used to explain the molecular-level effects that interfaces can have on these properties. When the same material components and conditions exist in composites, increases in E' are understood to be a function of stronger molecular pinning [1] (immobilisation of

Table 8.1. Factors that influence the strength of interfacial interactions in reinforced plastics.

Factors	Additional notes
Reinforcement type	– Particulates – Long fibres – Short fibres – Whiskers – Nanoparticles
Available surface area	– A function of reinforcement size and shape – Should be related to the overall volume availability – Available area is of relevance when the reinforcement surface is fully 'wet' by the matrix
Reinforcement distribution	– Particle or fibre size distribution can be relevant – Closeness of distributed fibres or particles can affect molecular structures and interactions of interphases – Distribution and closeness can be intimately tied to reinforcement volume fraction – Use of dispersing agents can affect things like homogeneity and fibre/particle agglomeration
Reinforcement shape	– Fibre/particle/whisker geometry can affect the availability of the surfaces and molecular arrangements
Surface chemistry	– Affected by chemical or physical treatment to reinforcement prior to inclusion within matrix – Affected by type of sizing used on fibre surfaces – Different types of reinforcements can have original surface chemistries that are both anionic and cationic (e.g. kaolins) and may require chemical dispersing

molecules) at interfaces, and thus stronger adhesion between reinforcement and matrix. What is often also noticeable from a vast range of publications dealing with adhesion topics in relation to E' and E'' is that E'' is also reported to increase with improved adhesion [10, 11]. This can seem as counter intuitive, since the E'' indicates energy loss and since more energy is stored, one would think less energy would be lost. Since in the glassy region, molecules are essentially locked, motion force through loading occurs with greater difficulty and thus more energy can be stored. However, this also means more energy is required to onset friction through molecular motion, which dissipates the force to heat, which results in an increase of E''.

8.3.2 Changes to the tan δ (loss factor) curve

There are numerous notable variations in the tan δ curve that can be associated with changes to the strength of bonding at reinforcement interfaces, each of which is

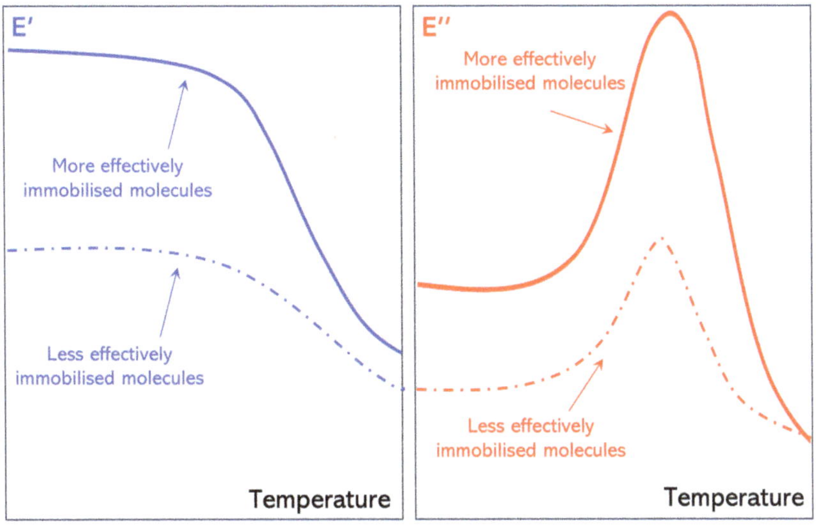

Figure 8.4. Idealised E' and E'' curves often observed in studies focussed on comparing the effects of reinforcement surface treatments in composite materials.

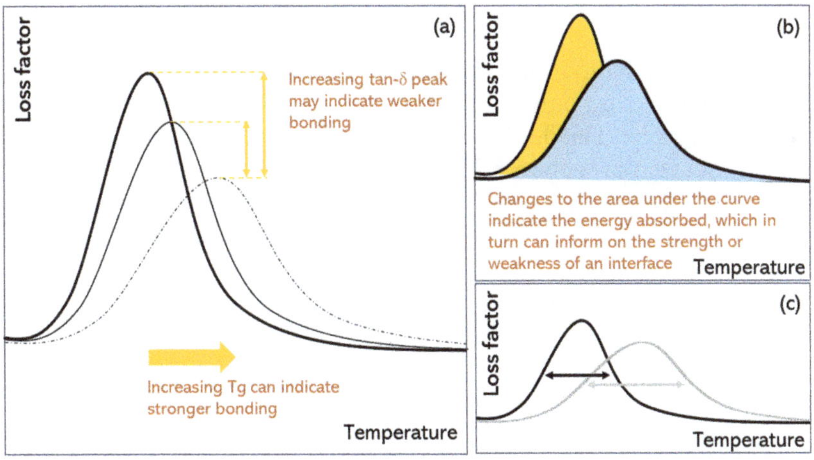

Figure 8.5. Idealised tan δ (loss factor) curves often observed in studies focussed on comparing the effects of reinforcement surface treatments in composite materials assuming the bonding at interfaces is the only parameter that differs between examples: (a) stronger interfacial adhesion can be inferred from increases to Tg and decreases in the tan δ, (b) differences in molecular pinning at reinforcement interfaces can affect the damping properties of composites, and (c) increased or decreased heterogeneity in a composite can be a function of interfaces and the polymer phases that result from the way in which molecules pin to interfaces and the resultant effects on interphases. Changes to the heterogeneity from interfaces can be noticed through direct measurements of the width of the tan δ curve, usually at half of the peak height.

illustrated in figure 8.5. The Tg is one example, which if taken from the tan δ peak, should increase as a function of higher reinforcement bonding strength. Since strong adhesion typically results in reduced molecular mobility at interfaces, this should show up as an increase in the Tg [13]. Concurrently, since more weakly bonded interfaces have higher levels of molecular mobility, this results in higher levels of energy dissipation by internal friction and thus a higher tan δ peak [14–17], figure 8.5 (a). The area under the tan δ curve can also be an indicator of variations owing to changes in the strength of adhesion at the reinforcement interfaces [18]. This is because pinned molecules can be either weakly or strongly pinned to interfaces, resulting in differences to the dynamic energy that can be absorbed. These are variations that can be measured directly via a temperature sweep, figure 8.5(b). Importantly, molecular arrangements at interfaces can therefore affect the damping capacity of a composite material. Finally, variations in the width of the tan δ can indicate introduced or reduced heterogeneity in a composite material as a function of molecular interactions at reinforcement interfaces and the effects these have on connected interphases. This is measurable via the width of the tan δ curve, which is most typically taken at half of the peak height of the curve [18, 19].

8.3.3 Changes within the rubbery region

Fibre surface pretreatment can affect localised crosslinking and molecular entanglements. This can show up in both E' and E'' modulus curves as shown in figure 8.6. When fibre pretreatment results in increased crosslinking and/or entanglements, the E' and E'' curves appear higher relative to composites containing the same material components, but which differ in terms of fibre pretreatment [20]. As such, the effects of composite interfaces on polymer properties are noticeable under controllable and comparable conditions. The degree of entanglements can in fact be calculated using E', which is discussed in section 8.4.2.

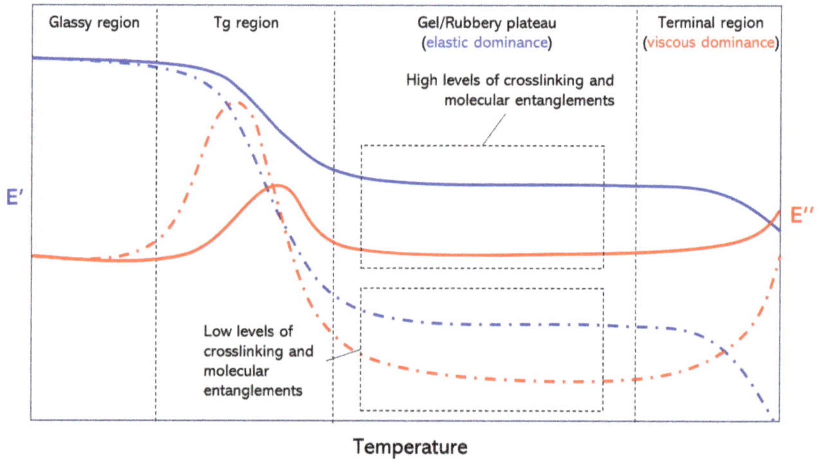

Figure 8.6. Idealised E' and E'' curves of composite materials with differing fibre pretreatment in a temperature sweep. Expected positions of each curve given higher and lower levels of crosslinking and molecular entanglement are shown.

8.4 Useful DMA models

8.4.1 The adhesion factor, *A*

The adhesion factor, A, equation (8.9), [12] is useful for determining the relative adhesion of reinforcement to polymer in a dynamically loaded composite. Here, V_F is the volume fraction of reinforcing fibre in the composite, and $\tan\delta_c(T)$ and $\tan\delta_m(T)$ are the loss factors at temperature T for the reinforced composite and the matrix materials, respectively. The value of A is inversely related to reinforcement-matrix adhesion [13] such that lower values of A indicate higher levels of adhesion (i.e. greater interaction between the reinforcements and the polymer matrix). Similarly, higher values of A indicate lower levels of reinforcement-matrix adhesion (i.e. reduced interaction between the reinforcements and the polymer matrix).

$$A = \frac{1}{1 - V_f}\frac{\tan\delta_c(T)}{\tan\delta_m(T)} - 1 \tag{8.9}$$

8.4.2 The degree of entanglement, *N*

Surface treatments to reinforcement can also affect the degree of entanglement, which is hence another useful comparative parameter that can be ascertained using DMA methods. The degree of entanglement, N (units mol $\cdot$ m^{-3}), equation (8.10), is calculated using the composite storage modulus taken from the plateau region of the E' versus temperature curve, the universal gas constant, R, and the absolute temperature, T [23]. Higher values of N indicate higher levels of entanglement due to the presence of interfaces. Since cohesive forces are primarily responsible for surface tension, higher entanglement also indicates improved adhesion and reduced interfacial tension.

$$N = \frac{E'}{6RT} \tag{8.10}$$

8.4.3 The reinforcing effectiveness factor, *C*

The effectiveness of reinforcement embedded into a polymer matrix can be determined using a reinforcing effectiveness factor, C, equation (8.11), which is applicable for a variety of polymer composites [21, 22]. Here, E_g' and E_r' represent the storage moduli values in the glassy and rubbery regions, respectively. Both glassy to rubbery storage moduli are expressed as a linear ratio for both the reinforced composite and for the neat unreinforced polymer. Since these values do vary as a function of temperature within both glassy and rubbery regions, the better practice is to stipulate the precise temperature at which the individual values are taken. The value of C informs on the reinforcing effectiveness within the polymer. Similarly to the adhesion factor, the reinforcing effectiveness factor is inversely related to the strength of adhesion, with higher values of C indicating less effective reinforcing, and lower values of C indicating more effective reinforcing.

$$C = \frac{\left[\dfrac{E_g'}{E_r'}\right]_{\text{composite}}}{\left[\dfrac{E_g'}{E_r'}\right]_{\text{polymer}}} \tag{8.11}$$

References

[1] Touaiti F, Alam P, Toivakka M and Bousfield D 2010 Polymer chain pinning at interfaces in $CaCO_3$–SBR latex composites *Mater. Sci. Eng.* A **527** 2363–9

[2] Touaiti F, Pahlevan M, Nilsson R, Alam P, Toivakka M, Bousfield D and Wilen C E 2013 Impact of functionalised dispersing agents on the mechanical and viscoelastic properties of pigment coating *Prog. Org. Coat.* **76** 101–6

[3] Bashir M A 2021 Use of dynamic mechanical analysis (DMA) for characterizing interfacial interactions in filled polymers *Solids* **2** 108–20

[4] Nwambu C N, Robert C and Alam P 2022 Viscoelastic properties of bioinspired asymmetric helicoidal CFRP composites *MRS Adv.* **7** 805–10

[5] Qiao J, Zhang Q, Wu C, Wu G and Li L 2022 Effects of fiber volume fraction and length on the mechanical properties of milled glass fiber/polyurea composites *Polymers* **14** 3080

[6] Mittal M and Chaudhary R 2018 Effect of fiber content on thermal behavior and viscoelastic properties of PALF/Epoxy and COIR/Epoxy composites *Mater. Res. Exp.* **5** 125305

[7] Vacche S D, Oliveira F, Leterrier Y *et al* 2012 The effect of processing conditions on the morphology, thermomechanical, dielectric, and piezoelectric properties of P(VDF-TrFE)/ $BaTiO_3$ composites *J. Mater. Sci.* **47** 4763–74

[8] Dalmas F, Cavaille J Y, Gauthier C, Chazeau L and Dendieval R 2007 Viscoelastic behavior and electrical properties of flexible nanofiber filled polymer nanocomposites. Influence of processing conditions *Compos. Sci. Technol.* **67** 829–39

[9] Nwambu C N, Robert C and Alam P 2022 Dynamic mechanical thermal analysis of unaged and hygrothermally aged discontinuous Bouligand structured CFRP composites *Funct. Compos. Struct.* **4** 045001

[10] Mylsamy K and Rajendran I 2011 Influence of alkali treatment and fibre length on mechanical properties of short Agave fibre reinforced epoxy composites *Mater. Des.* **32** 4629–40

[11] Yu T, Ren J, Li S, Yuan H and Li Y 2010 Effect of fiber surface-treatments on the properties of poly(lactic acid)/ramie composites *Composites A* **41** 499–505

[12] Jyoti J, Singh B P, Arya A K and Dhakate S R 2016 Dynamic mechanical properties of multiwall carbon nanotube reinforced ABS composites and their correlation with entanglement density, adhesion, reinforcement and C factor *RSC Adv.* **6** 3997

[13] Goriparthi B K, Suman K N S and Rao N M 2012 Effect of fiber surface treatments on mechanical and abrasive wear performance of polylactide/jute composites *Composites A* **43** 1800–8

[14] Keusch S and Haessler R 1999 Influence of surface treatment of glass fibres on the dynamic mechanical properties of epoxy resin composites *Composites A* **30** 997–1002

[15] Liu Y, Zhang X, Song C, Zhang Y, Fang Y, Yang B and Wang X 2015 An effective surface modification of carbon fiber for improving the interfacial adhesion of polypropylene composite *Mater. Des.* **88** 810–9

[16] Fang L, Chang L, Guo W J, Chen Y and Wang Z 2014 Influence of silane surface modification of veneer on interfacial adhesion of wood-plastic plywood *Appl. Surf. Sci.* **288** 682–9
[17] Joshy M K, Mathew L and Joseph R 2007 Studies on interfacial adhesion in unidirectional isora fibrereinforced polyester composites *Compos. Interfaces* **14** 631–46
[18] Karakaya N, Papila M and Ozkoc G 2020 Effects of hot melt adhesives on the interfacial properties of overmolded hybrid structures of polyamide-6 on continuous carbon fiber/epoxy composites *Composites A* **139** 106106
[19] Liao M, Yang Y and Hamada H 2016 Mechanical performance of glass woven fabric composite: effect of different surface treatment agents *Composites B* **86** 17–26
[20] Haris N I N, Hassan M Z, Ilyas R A, Suhot M A, Sapuan S M, Dolah R, Mohammad R and Asyraf M R M 2022 Dynamic mechanical properties of natural fiber reinforced hybrid polymer composites: a review *J. Mater. Res. Technol.* **19** 167–82
[21] Pothan L A, Oommen Z and Thomas S 2003 Dynamic mechanical analysis of banana fiber reinforced polyester composites *Compos. Sci. Technol.* **63** 283–93
[22] Bagotia N and Sharma D K 2019 Systematic study of dynamic mechanical and thermal properties of multiwalled carbon nanotube reinforced polycarbonate/ethylene methyl acrylate nanocomposites *Polym. Test.* **73** 425–32
[23] Oommen Z, Groeninckx G and Thomas S 2000 Dynamic mechanical and thermal properties of physically compatibilized natural rubber/poly(methyl methacrylate) blends by the addition of natural rubber-graft-poly(methyl methacrylate) *J. Polym. Sci.* B **38** 525–36

IOP Publishing

Composite Interfaces in Mechanical Design

Parvez Alam

Chapter 9

Fracture and failure at interfaces in composites

9.1 Introduction

This chapter will consider specific details on fracture modes and failure behaviours at composite interfaces of continuous fibre-reinforced plastics. To start the chapter we develop a broad understanding of overall composite behaviour in terms of load and deformation, as this informs the mode of loading an interface may be subjected to, which in turn influences both failure and fracture characteristics at interfaces. The chapter then continues to discuss failure at both micro- and macrostructural levels, highlighting the effects of the loading mode on fracture and failure and touching on the effects of environmental conditioning.

9.2 Tensile loading and fibre orientation

Tensile loading is the most common loading condition used in research to characterise the properties of fibre-reinforced composites. It is clear that tensile loading should be considered relative to the fibre direction of the composite in continuous fibre-reinforced plastics. Figure 9.1(a) provides schematic representations of tensile stress–strain curves for continuous fibre-reinforced plastics where the load is applied at 0°, 15°, 45°, and 90° relative to the 0° fibre axis. It can be noted that composite strength typically decreases as a function of increasing fibre orientation away from the loading axis, as schematised in figure 9.1(b). The 0° orientation is by far the stiffest and strongest fibre orientation relative to the loading axis, which is because when oriented at 0°, the fibres play a significant role in load bearing such that the Rule of Mixtures can be applied to predict both tensile modulus, E_t, equation (9.1), and tensile strength, σ_t, equation (9.2), such that the reinforcement is assumed to be used to its fullest extent. In these equations, V_m is the volume fraction of matrix, V_f is the volume fraction of fibre, E_m is the tensile modulus of matrix polymer, E_f is the tensile modulus of fibre reinforcement, σ_m is the tensile strength of matrix polymer, and σ_f is the tensile strength of fibre reinforcement. Loading at $>0°$ results in a reduction in both modulus and strength values, but an increase in the elongation to failure.

doi:10.1088/978-0-7503-5688-6ch9

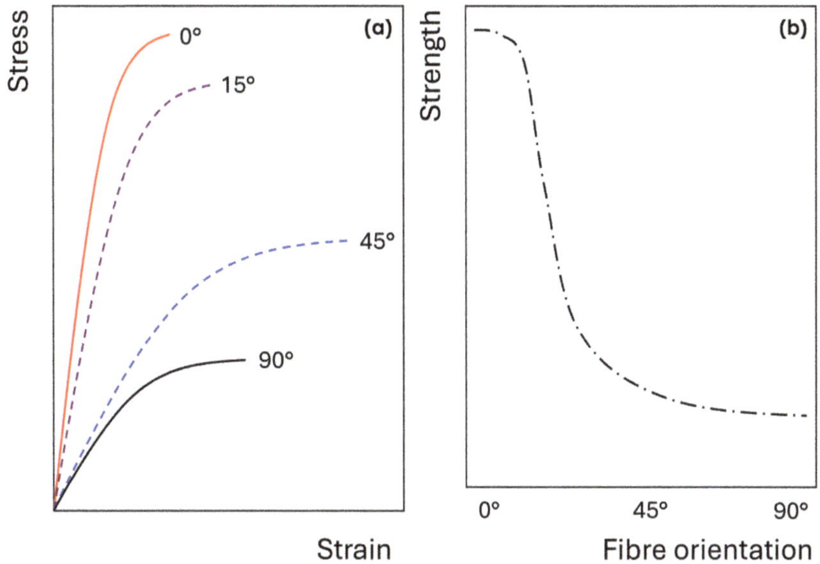

Figure 9.1. (a) Schematic representations of stress–strain curves for continuous fibre-reinforced plastics (unidirectionally aligned) with static loading applied parallel to the fibre axis (i.e. 0°), at 15° off the fibre axis, at 45° off the fibre axis, and perpendicular to the fibre axis (i.e. 90°), and (b) representative curve for strength plotted against the fibre orientation.

The drop in modulus and strength due to off-axis fibre orientation can be crudely approximated in a single dimension, using trigonometric principles, equations (9.3) and (9.4), where θ is the fibre orientation angle in radians. While this single dimensional approximation should not be considered a replacement for a complete composite laminate theory calculation, it does provide simple and quick reasoning for why the reinforcement orientation will have such a significant effect on composite strength and stiffness. The elongation to failure is notably highest at 45°, which is due to the change from fibre-dominated properties to matrix--dominated properties.

The strain to failure decreases from 45° to 90°, which is related to the matrix yielding at interfaces through shear, which is in turn influenced by the constraints on fibres and the effective length of material actively shearing. Fibre constraints are lowest in the 45° orientation, and this provides the greatest freedom for shear yielding at fibre–matrix interfaces. As the orientation of the fibres changes from 45° to 90°, the length of effective material in shear, and therefore the effective yielding of interface, decreases as the loading mode changes from being dominated by interfacial shear, to one where the pull-off strength of the interface dominates failure, and the modulus and strength are calculated using a Transverse Rule of Mixtures approach, equations (9.5) and (9.6). The Rule of Mixtures and the Transverse Rule of Mixtures are typically deemed to be the upper and lower bounds, respectively, describing composite properties, figure 9.2.

$$E_t = V_m E_m + V_f E_f \tag{9.1}$$

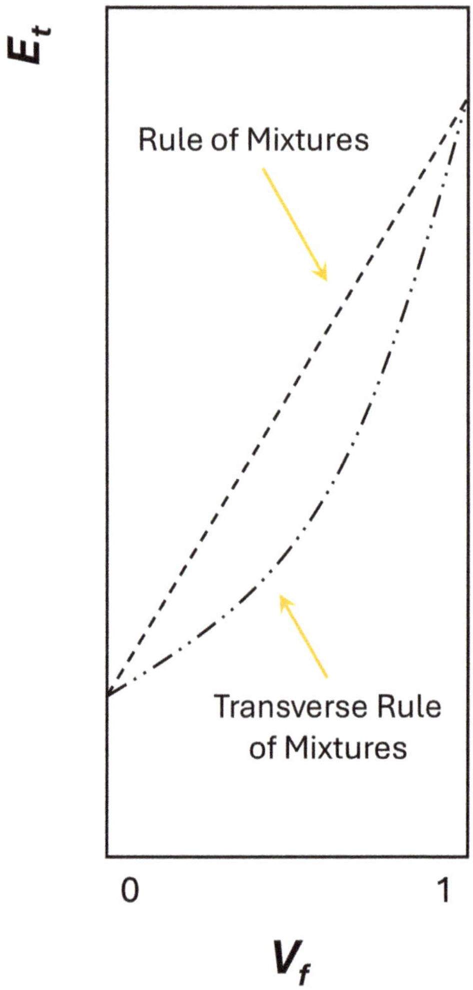

Figure 9.2. The Rule of Mixtures model and the Transverse Rule of Mixtures model as upper and lower limits describing composite properties.

$$\sigma_t = V_m \sigma_m + V_f \sigma_f \tag{9.2}$$

$$E_t(\theta) = V_m E_m + V_f \cos \theta E_f \tag{9.3}$$

$$\sigma_t(\theta) = V_m \sigma_m + V_f \cos \theta \sigma_f \tag{9.4}$$

$$E_t = \left(\frac{V_f}{E_f} + \frac{V_m}{E_m} \right)^{-1} \tag{9.5}$$

$$\sigma_t = \left(\frac{V_f}{\sigma_f} + \frac{V_m}{\sigma_m}\right)^{-1} \tag{9.6}$$

The differences in loading mode as a function of fibre orientation are illustrated in a simplified manner in figure 9.3. While this figure provides information on the primary types of loading at fibre interfaces, other factors will have an additional, albeit less pronounced influence on the mode of loading at fibre–matrix interfaces. One example is the thinning of material transverse to the loading direction. The extent of thinning differs between composite components since the Poisson's ratio of fibre and matrix is typically not the same. In addition, the varied levels of strain they experience under the same loading will also affect the Poisson's effect in individual components. For example, in a 0° continuous fibre-reinforced composite, this will induce minor tensile (pull off type) forces at fibre interfaces transverse to fibre axis. It should nevertheless be noted that the primary modes of loading at interfaces are typically the ones that will have the greatest influence over how the interface reacts and fails.

At the macroscale, the in-plane Poisson's ratio will typically decrease as the fibre orientation is taken further out of pitch with the angle of loading on the composite. This is shown in figure 9.4, which schematically illustrates the way in which the Poisson's ratio of continuous fibre composites varies as loading is applied at different angles to the fibre direction. Some physical reasoning behind this can be derived through the relationship $G = \frac{E}{2(1+\nu)}$, where it is clear that $G \propto \nu^{-1}$, i.e. the

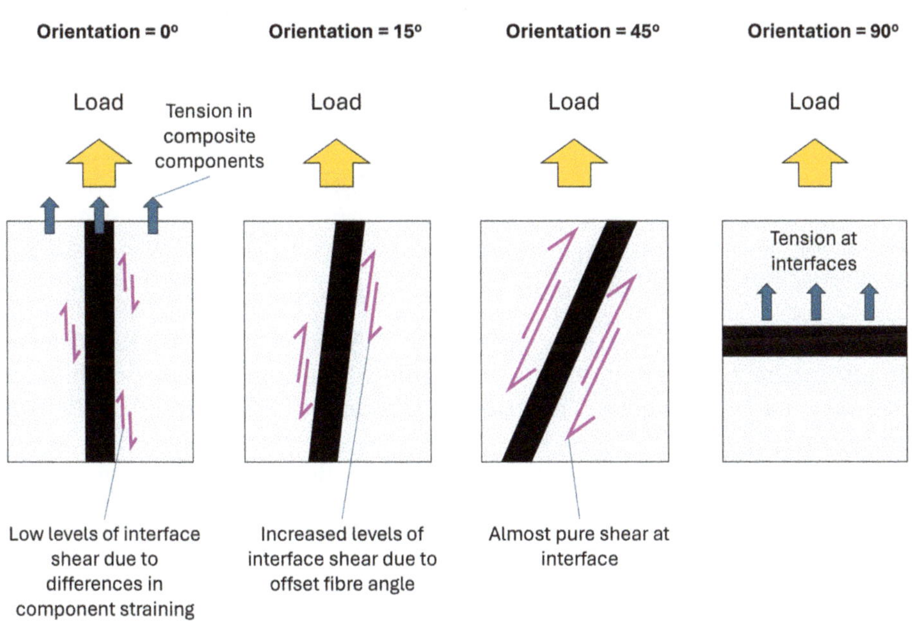

Figure 9.3. Schematic representation of (fundamental) different modes of loading experienced at fibre–matrix interfaces with respect to fibre orientation.

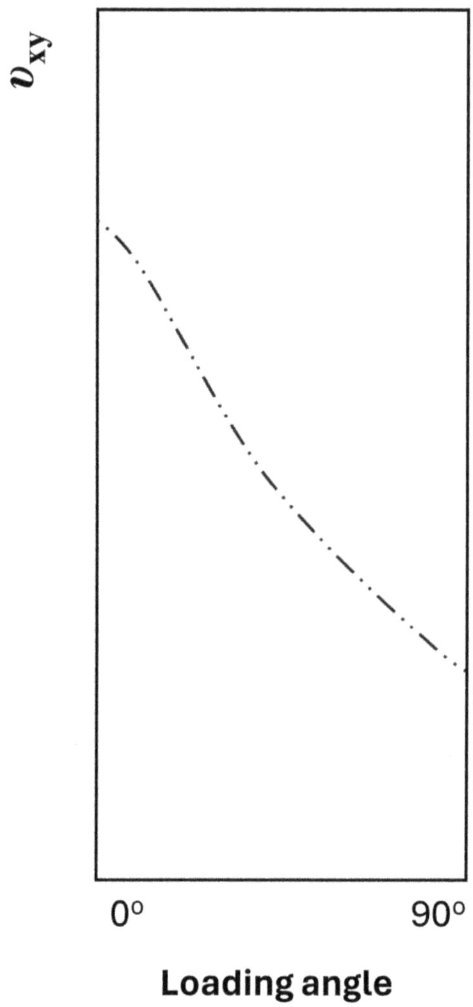

Figure 9.4. Schematic illustration of the variation of Poisson's ratio when plotted against the angle of applied load for unidirectional composites.

dependence of ν on G is inversely proportional. As local tensile loading decreases with a concurrent increase in local shear loading, the proportional contribution of elastic shear resistance increases relative to elastic tensile resistance and the Poisson's ratio is lowered.

When continuous fibre composites are laminated, the effects of loading angle and orientation remain unless the interfaces of laminates are constrained by neighbouring laminates laid at different orientation angles. For example, when cross-laminated neighbouring plies are laid atop one another, they exhibit mirrored orientations, and as such, adjacent plies will restrain the extent to which neighbouring plies can deform through their interfaces. Figure 9.5 shows representative stress–strain curves illustrated schematically for balanced [0/90] and [±45] continuous

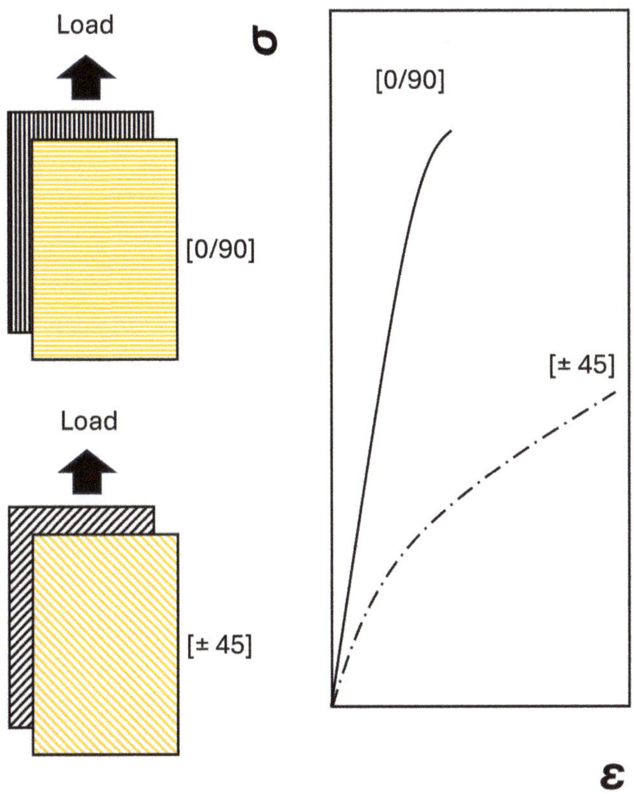

Figure 9.5. Representative stress–strain curves illustrated schematically for [0/90] and [±45] continuous fibre-reinforced plastic composites.

fibre-reinforced plastics. The expectation, as is clear from the figure, is that [0/90] laminates will have both higher stiffness and strength than [±45] laminates. Fundamentally, this is because half, or close to half, of the [0/90] laminates are unidirectionally oriented in the loading direction as such the fibre reinforcements bear load. Since the fibres have high strength and stiffness, very little of the matrix is involved and the stiffness and strength of such composites is higher than for [±45]-laminated composites. In the case of [±45]-laminated composites, due to the 45° fibre offset from the loading direction, the laminates experience shearing at the microstructural level and much of the load is borne by shear resistance in the matrix and at the fibre interfaces.

There are different effects on the Poisson's ratio that arise when loading balanced [0/90] and [±45] laminates at different angles. The Poisson's ratio of balanced [0/90] laminates is typically around 0.3, and as loading is applied at different angles to this, the Poisson's ratio rises as the fibre orientation relative to the loading direction enables greater movement in the matrix component of the composite. This reaches an apex when the loading angle is at 45° to the original [0/90] axis and shearing is high; the matrix component can stretch and fibres rotate in such a way that the

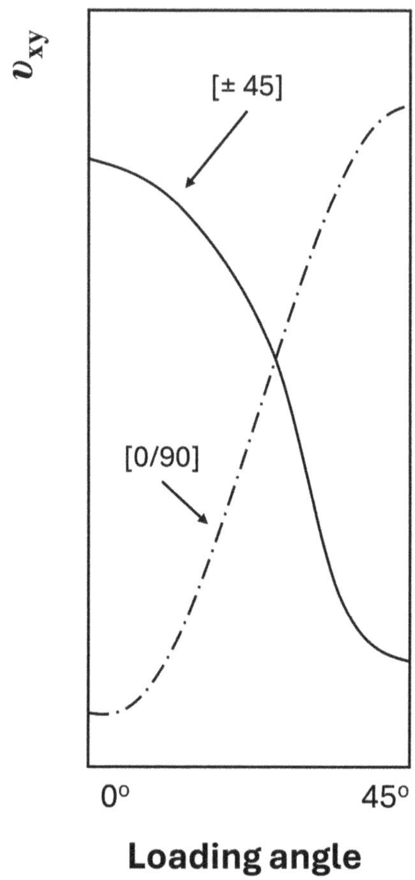

Figure 9.6. Schematically illustrated curves showing the variation of Poisson's ratio as a function of loading angle for [0/90] and [±45] continuous fibre-reinforced plastic composites.

transverse strain becomes very high relative to the axial strain, sometimes reaching up to 80% of the axial strain. The opposite occurs in balanced [±45] laminates which are already in a very high transverse strain state at a 0° loading angle. As the loading angle is increased, the ability for matrix material to stretch decreases, as does the ability for fibre to rotate. Consequently, the transverse strain decreases, and at a loading angle of 45°, the composite is effectively loaded as a [0/90] composite with a relatively lower ($\nu = 0.3$) Poisson's ratio (figure 9.6).

9.3 Tensile failure at the microstructural level

The properties of the fibre can be determined using single fibre tensile tests (SFTTs). Standards for SFTT are available such as ASTM D3822 [1], which focusses on measuring the tensile properties of both natural and manmade fibres; ISO 11 566 [2] and ASTM C1557 [3], which describe a method for determining the tensile properties of single-filament carbon fibres; and ASTM D2343-17 [4], which provides a method for the determination of the tensile properties of glass fibre strands, yarns, and rovings.

A generic example of an SFTT setup is shown in figure 9.7 where a fibre is essentially connected to two separated rigid mounts that when pulled apart tension the fibre. An SFTT is usually taken to failure, and taking ISO 11 566 [2] as an example from which fibre properties can be derived, the fibre tensile modulus, E_f, is calculated according to equation (9.7), where ΔF is the difference in force corresponding to the load limits of 400 mN/tex and 800 mN/tex (or the force corresponding to the strain limits selected, depending on the nominal strain at break of the fibre); A is the fibre cross-sectional area; L is the gauge length; ΔL is the difference in length at the load limits of 400 mN/tex and 800 mN/tex (or the difference in length corresponding to the strain limits selected, depending on the nominal strain at break of the fibre); and K is the system compliance.

$$E_f = \frac{\left(\frac{\Delta F}{A}\right)\left(\frac{L}{\Delta L}\right)}{1 - K\left(\frac{\Delta F}{\Delta L}\right)} \tag{9.7}$$

At the microstructural level, failure in tension occurs in the fibre, in the matrix, or at the fibre/matrix interface.

9.3.1 Fibre failure in tension

The type of fibre failure observed in static tension will depend on the properties of the fibre and in some cases on the rate of testing, figure 9.8. Carbon and glass typically exhibit brittle failure mechanisms, as do high crystallinity polymer fibres. Polymer fibres of lower crystallinity may neck and fail in a ductile manner. If the polymer is nevertheless viscoelastic, its behaviour will be strain rate dependent, and these fibres exhibit a ductile to brittle failure transition as a function of strain rate.

Brittle fibre failure modes exhibit Griffith brittle fractures, the fracture stress, σ_{fracture}, which is defined by equation (9.8), where E is the Young's modulus of the fibre, γ_s defines surface energy per unit area of surface created, and a is the crack length. Figure 9.9 from [5] shows different material examples of brittle fibre failure. Figure 9.9(a) shows a glass fibre with a smooth fracture that traverses the fibre

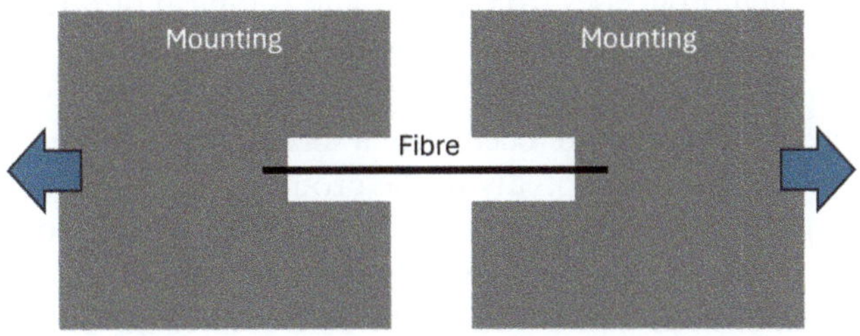

Figure 9.7. Generic setup of a SFTT consisting of a fibre attached to two mounting plates that are subjected to tension.

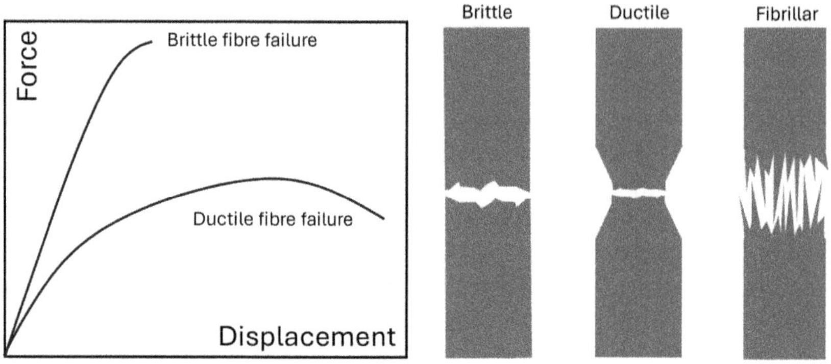

Figure 9.8. Reinforcing fibres: brittle, ductile, and fibrillar failure modes.

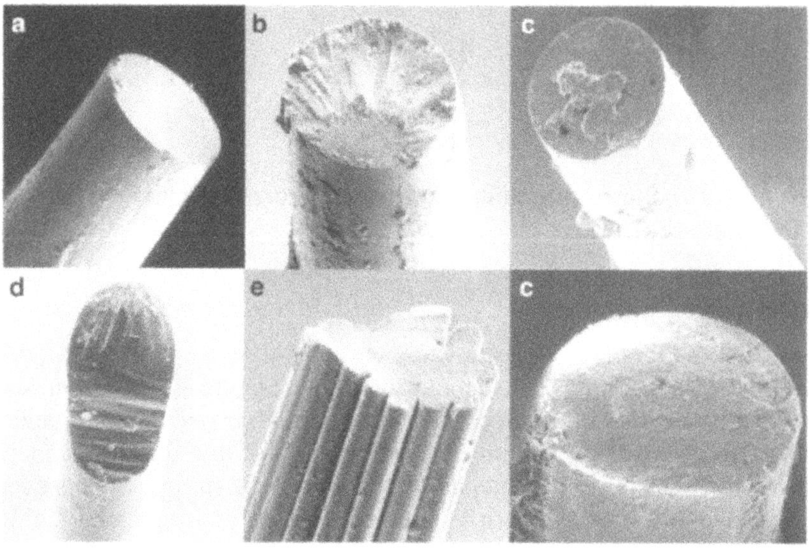

Figure 9.9. Brittle fractures. (a and b) Glass fibres. (c and d) Ceramic fibres, Nextel, Nicolon. (e) Carbon fibre from rayon precursor. (f) Elastomeric Lycra. Reprinted from [5]. Copyright (2002), with permission from Elsevier.

cross-section, the mirror region of which will often display multiple cracks leading to a hackle region, which progresses from a flaw and is typical of brittle materials with no yielding mechanisms, figure 9.9(b). Figures 9.9(c)–(d) show similar surfaces in brittle ceramic fibres (Nextile, Nicolon) and in figure 9.9(e) a smooth cross-sectional Griffith fracture can be seen to occur on a brittle carbon fibre from a rayon precursor. Figure 9.9(f) is the anomaly of the group as it is a polymer (Lycra). This polymer has a high strain to failure (>500%), which in typical polymers results in mechanisms such as ductile necking; however, the elastic modulus of Lycra increases as a function of extension, and at the point of fracture, the strain energy in the material is high and the fibre breaks as a Griffith solid with a smooth, unperturbed surface.

$$\sigma_{\text{fracture}} = \sqrt{\frac{2E\gamma_s}{\pi a}} \tag{9.8}$$

As mentioned, the more common failure mode in polymer fibres is ductile, as shown in figure 9.10. Ductile failure stress, σ_{failure}, is inadequately described by the classical Griffith equation, and is better described using an Irwin–Orowan modification of Griffith's equation, as shown in equation (9.9), where γ_p defines the plastic work per unit area of the surface created and the assumption is the $\gamma_p \gg \gamma_s$. Drawing affects the ductile failure modes of polymer fibres. Thick undrawn monofilament fibres such as the nylon fibres shown in figures 9.10(a) and (b) realise crack initiation from the void coalescence (a), which results in an increase of stress on the unbroken side of the fibre causing further plasticity and yielding at the crack. When such cracks open to a critical size and reach a critical stress level, catastrophic failure can occur, as shown in the polyester fibre in figure 9.10(c). Shear bands form around ductile yielding crack tips as shown using a polyester film in figure 9.10(d). The position and orientation of the initial crack line can alter the final fracture profile as shown in the nylon fibre in figure 9.10(e). These initial crack lines usually initiate at the fibre surface; however, internal flaws can cause ductile fibre failure, which is recognisable as it appears in the form of a double cone, as shown in figure 9.10(f).

$$\sigma_{\text{failure}} = \sqrt{\frac{2E(\gamma_s + \gamma_p)}{\pi a}} \tag{9.9}$$

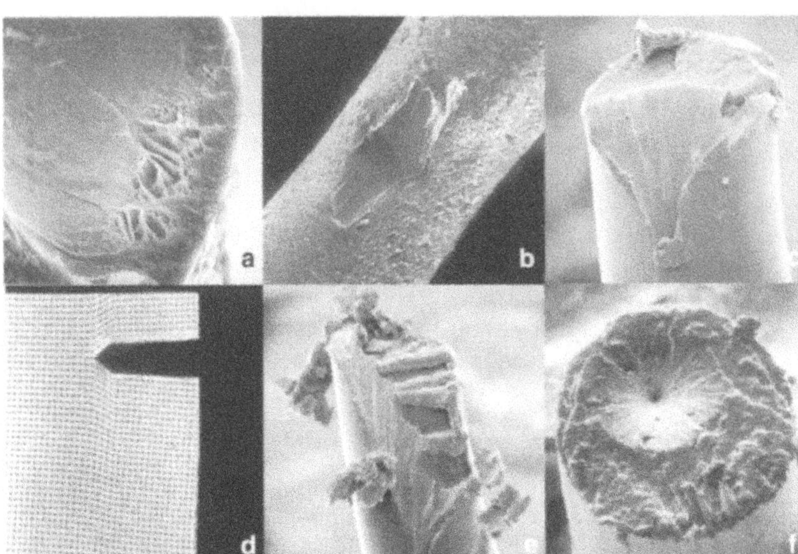

Figure 9.10. Ductile failures/fractures. (a and b) Thick undrawn nylon monofilament. (c) Polyester fibre. (d) Polyester film. (e and f) Nylon fibre. Reprinted from [5]. Copyright (2002), with permission from Elsevier.

A third commonly observed tensile fibre failure mode is fibrillar fracture. In this failure mode, the fibre splays unevenly, which is in part related to molecular/nanostructural organisation within the fibre itself. Fibre microstructures that are either made up of fibrils (such as natural fibres) or that are highly unidirectionally oriented at the molecular level with high molecular weight molecules and thus long crystalline molecular chains (such as pararamids) are prone to fibrillar failures. These can be visualised in figure 9.11 where in (a) a natural fibrillar fibre such as cotton fails in this way as absorbed water between the individual fibrils inhibits stress transfer, thus resulting in fibrillar failure, and in (b) and (c) where we see the highly oriented polymer Kevlar experiencing multiple splits and single splits, respectively. This occurs fundamentally when the intramolecular strength in the axis of loading far surpasses the intermolecular strength such that shear stresses lead to crack growth along the axis in what appears to be a fibrillar form of failure.

It is generally harder to predict the fracture stress, σ_{fracture}, for fibrillar failures, but the Griffith equation can be modified to enable closer predictions of failure, equation (9.10). In this generalised Griffith equation, ϕ essentially defines a fracture energy that corresponds to the brittle, plastic, viscoelastic, viscoplastic, crack meandering, and crack branching surface energies, which at the very least for fibrillar (shear dominated) fractures would need to take into consideration the nucleation of the intermolecular and/or interfibril slip pulses [6]. In a fibrillar failure the overall value of ϕ is difficult to predict, but in some simpler cases, an understanding of shear surface energies may suffice.

$$\sigma_{\text{fracture}} = \sqrt{\frac{2E\phi}{\pi a}} \tag{9.10}$$

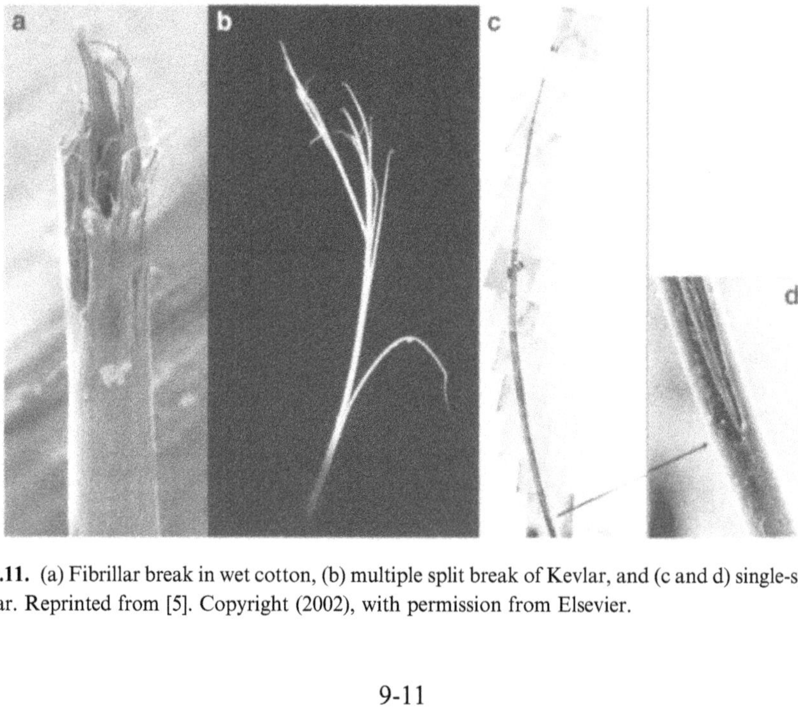

Figure 9.11. (a) Fibrillar break in wet cotton, (b) multiple split break of Kevlar, and (c and d) single-split break of Kevlar. Reprinted from [5]. Copyright (2002), with permission from Elsevier.

9.3.2 Matrix failure in tension

Failure can occur in the bulk of the matrix polymer in fibre-reinforced plastic composites. Various types of failure can be observed. Since the failures occur within bulk polymer and not at polymer interfaces, these failure modes are commonly observed in unreinforced polymer. Figures 9.12(a)–(f) provide images modified from [7] of different failure modes in polymer. A brittle failure mode is shown in figure 9.12(a) and is often a consequence of either a high cross-linking density in the polymer or long polymer molecules with strong (crystalline) intermolecular bonding. Each of these restricts the mobility of the polymer molecules under loading and thus the polymer will absorb strain energy until it reaches a critical limit for energy release, at which point the material will fracture. Figure 9.12(b) shows polymer micro-crazing, which develops when there are excess tensile stresses within the polymer. This leads to the formation of inhomogeneity-initiated voids perpendicular to the primary axis of applied stress. Micro-crazing is a post-yielding phenomenon that tends to happen more often in glassy/amorphous polymers and is a precursor stage to fracture. Figure 9.12(c) shows a case of necking and yielding in the polymer, which typically initiates from an instability in tension when there is a disproportionate rate of decrease in cross-sectional area relative to the rate at which the polymer can strain harden. Figure 9.12(d) shows semi-ductile cracking, which indicates that the material has surpassed its yield strength, but is concurrently experiencing a parallel failure mode such as the fibrillar-type failure seen in (d). This is physically and visually very different than a pure ductile failure as seen in figure 9.12(e). Finally high levels of plasticity (plastic straining) leading to a cross-fracture are shown in figure 9.12(f).

At a laminate level, matrix cracking is commonly observed in off-axis plies in cross-ply laminated composites. Figure 9.13 provides examples of off-axis matrix cracking in (a) [0/75]2s laminate and (b) [0/90]2s laminate CFRP prepreg of T700SC/2592 (Toray Industries, Inc.) [8], and (c) [0/90/0] and (d) [0/90/0]s laminates

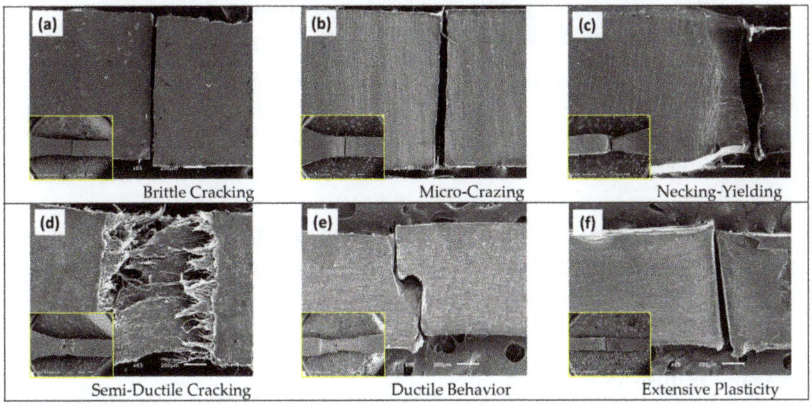

Figure 9.12. Example SEM images of polymer failure modes: (a) brittle cracking, (b) micro-crazing, (c) necking-yielding, (d) semi-ductile cracking, (e) ductile failure, and (f) plasticity. Reproduced from [7]. CC BY 4.0.

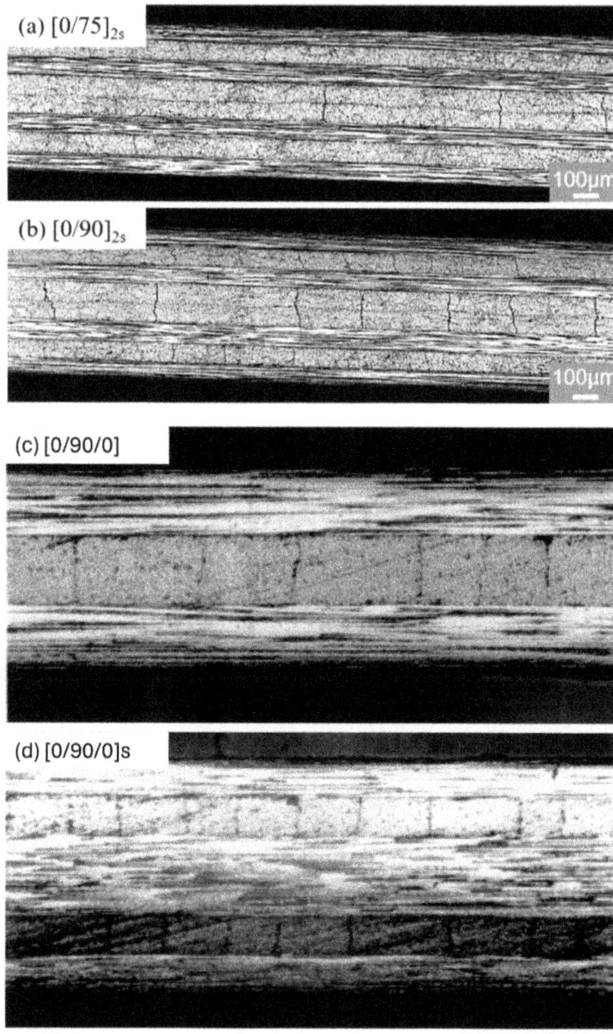

Figure 9.13. Example images of matrix cracking in off-axis plies of laminated composites: (a) [0/75]2s laminate and (b) [0/90]2s laminates CFRP prepreg of T700SC/2592 (Toray Industries, Inc.), (c) [0/90/0] and (d) [0/90/0]s laminates of AS4/3501-6 graphite–epoxy composite. (Panels (a) and (b) reproduced from [8], CC BY 4.0 and panels (c) and (d) reprinted from [9], Copyright (2000), with permission from Elsevier.)

of AS4/3501-6 graphite–epoxy composite [9]. Since matrix material will typically fail at lower loads than high load bearing reinforcing fibres, matrix cracking will be visible before the composite laminate fails as a whole.

9.3.3 Interface failure in tension

The mechanisms of bonding at fibre–matrix interfaces affect the detachment of fibres from matrices, both in terms of the force requirement for detachment, and in terms of the mode of detachment. These mechanisms of adhesion include mechanical interlocking, electrostatic adhesion, chemical adhesion, and diffusion adhesion, each of which has been covered in more detail in chapter 3. There are two main interfacial failures commonly observed through SEM. These include fibre–matrix debonding and fibre pullout.

9.3.3.1 Fibre debonding

Figure 9.14 shows scanning electron microscope (SEM) examples of post-loading fibre–matrix debonding from different perspectives with (A) showing a clean-cut cross-section from an E-glass/epoxy composite [10], (B) an angular view of debonding in a fractured basalt fibre/polypropylene composite [11], and (C) an angular view of debonding in a fractured E-glass/epoxy composite [12]. The evolution of in-plane strain as measured by digital image correlation (DIC) around a single fibre eventuating in debonding [13] is shown in (D) where (a) ε_{yy}, (b) ε_{xx}, and (c) ε_{xy} with global stress.

Numerous analytical models are used to predict stress at which fibre–matrix debonding occurs [14]. Shear lag-based models, equation (9.11), can be used to predict the debond stress, σ_d. In this model, shear strength at the interface, τ_0, is a function of the applied load, E_f is the Young's modulus of the fibre reinforcement, and G_m is the shear modulus of the matrix polymer. The function $\left(\ln\left(\frac{r_m}{r_f}\right)\right)^{\frac{1}{2}}$ is

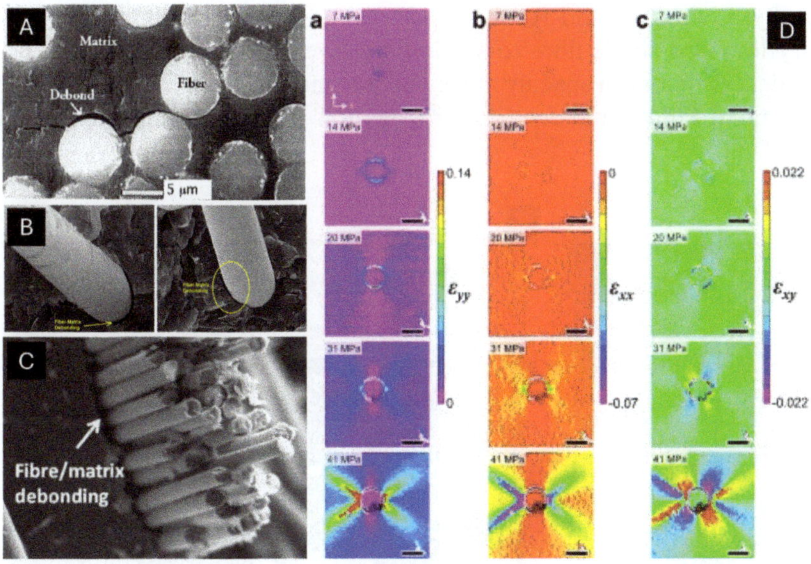

Figure 9.14. (A–C) SEM examples of post-loading fibre–matrix debonding in fibre-reinforced composites from different perspectives: (A) E-glass/epoxy resin, (B) basalt fibre/polypropylene, (C) E-glass/epoxy resin, and (D) dvolution of in-plane strain components as measured by digital image correlation (DIC) where (a) ε_{yy}, (b) ε_{xx}, and (c) ε_{xy} with global stress. Tensile load is applied in y-direction. All scale bars represent 2 mm. (A) Reprinted from [10]. Copyright (2019), with permission from Elsevier. (B) Reproduced from [11]. CC BY 4.0. (C) Reprinted from [12]. Copyright (2021), with permission from Elsevier. (D) Reprinted from [13]. Copyright (2023), with permission from Elsevier.

dependent on the fibre geometry where r_f is the fibre radius and r_m is the *effective* radius of matrix material surrounding the fibre. The effective matrix radius is difficult to predict and depends on the properties of the matrix, the fibre, and the surfaces (including sizing), since the effective matrix radius will to an extent also be correlated to the thickness of any interphase that exists. The Hashin-Rosen cylinder model [15] is oft-used in relationships such as is shown in equation (9.11) as it permits the derivation of closed-form expressions for effective properties [16].

$$\sigma_d = \tau_0\left(\frac{E_f}{2G_m}\right)^{\frac{1}{2}}\left(\ln\left(\frac{r_m}{r_f}\right)\right)^{\frac{1}{2}} \tag{9.11}$$

A drawback in using equation (9.11) is that it cannot take into account stress concentrations at the crack front at the fibre–matrix interface during debonding, and as such, may lead to overestimations of σ_d. As such, models incorporating the critical strain energy release rate for the interfacial debond may be superior. Equation (9.12) takes this into account as the critical strain energy release rate for mode II fracture, G_{IIc}, has notably first- order proportionality with σ_d.

$$\sigma_d = \left(\frac{4E_f G_{\text{IIc}}}{r_f}\right)^{\frac{1}{2}} \tag{9.12}$$

Recently, Livingston and Koohbar [19] demonstrated by DIC inter-fibre failure at a very localised single fibre to single fibre level, figures 9.15(A) and (B). Their work evidences that both spacing angle and distance between adjacent fibres plays a role in debond failure and fracture propagation between composite components,

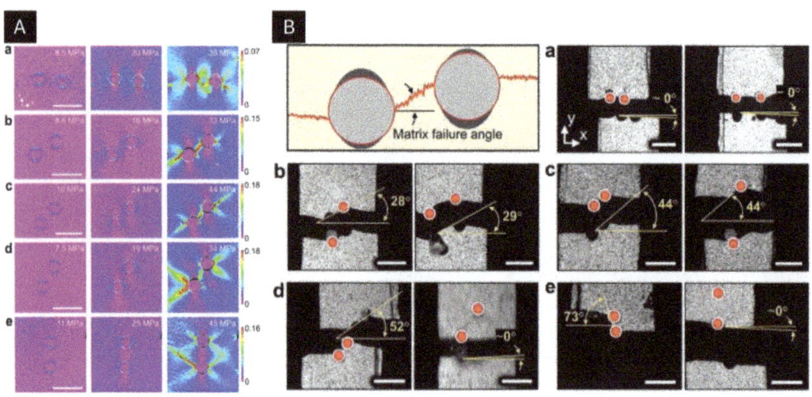

Figure 9.15. (A) DIC-measured evolution of longitudinal strain fields (ε_{yy}) at various global stresses and in the vicinity of double glass fibre samples: (a) 0°, (b) 30°, (c) 45°, (d) 60°, and (e) 90°. The contour maps on the right column are extracted at global stresses just before matrix failure in all cases. Scale bar: 5 mm. (B) Postmortem matrix failure angle measurements for double-fibre samples with short (left column) and long (right column) inter-fibre spacings. Data shown for samples with (a) 0°, (b) 30°, (c) 45°, (d) 60°, and (e) 90°. Red circles indicate the original locations of the fibres. Scale bar: 5 mm. Reprinted from [19]. CC BY NC-ND.

with fibre angle playing a more significant role in crack propagation than fibre spacing. In such cases, not only do fibres debond, but there is additional shear band coalescence, which is the cause of inter-fibre failure. In figure 9.15(A) we see the DIC-measured longitudinal strain fields (ε_{yy}) at various global stresses and in the vicinity of double glass fibre samples: (a) 0°, (b) 30°, (c) 45°, (d) 60°, and (e) 90°. The contour maps on the right column are extracted at global stresses just before matrix failure in all cases. In figure 9.15(B) we see the postmortem matrix failure angle measurements for double-fibre samples with short (left column) and long (right column) inter-fibre spacings. Data in this figure is shown for samples with (a) 0°, (b) 30°, (c) 45°, (d) 60°, and (e) 90°. Red circles indicate the original locations of the fibres.

9.3.3.2 Fibre pullout

Fibre pullout is a failure mode that occurs when either the shear bond strength or the frictional stress between fibre and matrix is surpassed. The failure mode is typically identified through observation of SEM images where a fibre pullout leaves either a visibly empty hole or a near semi-cylindrical cavity in the matrix component of the fracture surface. Examples of pullouts from these different perspectives are shown in figure 9.16, where in (a) a transverse to fibre axis view reveals semi-cylindrical cavities and in (b–d) fibre pullouts are recognised as holes left in the matrix visible at the fracture surface.

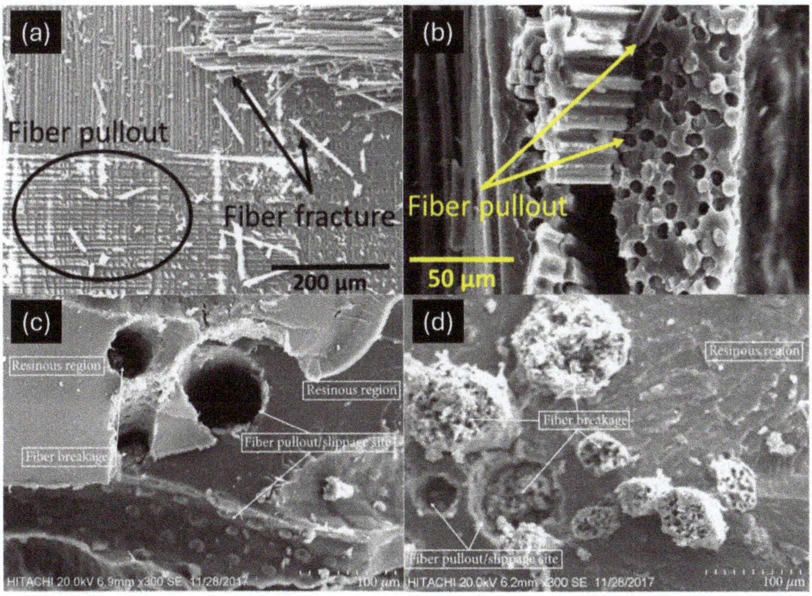

Figure 9.16. Aerospace-grade CYCOM® 937A plain weave, woven carbon fibre prepreg/epoxy (Solvay Industries Inc.) composite showing (a) transverse to fibre axis semi-cylindrical cavities due to pullout and (b) an angular view of a hole due to pullout. Sugar palm fibre-reinforced polyester composites (c–d) holes showing holes left in the matrix at the fracture surface due to pullout. Figures (a–b) reproduced from [20], CC BY 4.0, and (c–d) reproduced from [21], CC BY 4.0.

Shear bond strength is fundamentally linked to interfacial friction between fibre and matrix. Interfacial friction can arise through compressive radial stresses that are a function of resin shrinkage while curing or thermal mismatches between fibre and matrix during cooling. The frictive stresses as a function of distance, d, σ_{friction}, can be modelled using the non-linear relationship shown in equation (9.13), where $\sigma_p = \frac{\varepsilon_0 E_f}{\nu_f}$ and $\beta = \frac{2\mu\nu_f E_m}{E_f r_f(1+\nu_m)}$. Here, subscripts f and m refer to fibre and matrix, respectively; ν is Poisson's ratio; μ is the coefficient of friction between fibre and matrix; E is the Young's modulus; r_f the fibre radius; and ε_0 is a misfit strain between fibre and matrix. The non-linear relationship shown in equation (9.13) is generally seen as more accurate than linear relationships such as shown in equation (9.14), as this linear model ignores inevitable Poisson contractions. In this equation, $\tau_f = \mu P$ where P is an averaged compressive stress over the fibre surface [14].

$$\sigma_{\text{friction}}(d) = \sigma_p(1 - \sigma^{-\beta d}) \tag{9.13}$$

$$\sigma_{\text{friction}}(d) = \frac{2\tau_f d}{r_f} \tag{9.14}$$

The Kelly–Tyson formula [22], equation (9.15), has been a basis for a plethora of modified fibre–matrix shear stress, τ, models. Here, $\sigma_{f,b}$ is the fibre-breaking stress, D_f is the fibre diameter, and L_c is the critical transfer length. The Kelly–Tyson model took into consideration continuous fibres in a ductile matrix. These fibres experience multiple breakages and maintain load-bearing capacity until a critical length is reached. The model assumes perfectly plastic behaviour of the matrix, ideal adhesion between fibre and matrix, and a constant shear stress about the length of the embedded fibre.

$$\tau = \frac{\sigma_{f,b} D_f}{2L_c} \tag{9.15}$$

Drzal and Madhukar [23] note that there is in actuality a distribution of L_c and that Weibull statistics can be used to fit the data according to equation (9.16). The probability density function, $f(x)$, of the Weibull distribution is shown in equation (9.17), and the cumulative distribution, $F(x)$, of the Weibull distribution is shown in equation (9.18). In equations (9.16) to (9.18), α is a scaling parameter that affects the distribution scale, β is a shape parameter affecting the shape of the distribution, and the Γ function calculated as a function of β is $\Gamma(\beta) = \int_0^\infty \chi^{\beta-1} e^{-x} dx$.

$$\tau = \left(\frac{\sigma_{f,b}}{2\beta}\right)\Gamma\left(1 - \frac{1}{\alpha}\right) \tag{9.16}$$

$$f(x) = \frac{\alpha}{\beta}\left(\frac{x}{\beta}\right)^{\alpha-1} e^{-\left(\frac{x}{\beta}\right)^{\alpha}} \tag{9.17}$$

$$F(x) = 1 - e^{-\left(\frac{x}{\beta}\right)^{\alpha}} \tag{9.18}$$

Bader and Bowyer's [24, 25] proposed modification to the Kelly–Tyson model assumed that L_c exists at any value of composite strain, ε_c, and τ is represented by equation (9.19), where L_ε and E_f is the fibre modulus. As such, a Kelly–Tyson stress level is assumed in fibres with a length $<L_\varepsilon$, but when the fibre length $>L_\varepsilon$, the average fibre stress, σ_f, follows equation (9.20), and the composite stress, σ_c, is calculated according to equation (9.21), where i is an index referring to fibres where $L < L_\varepsilon$, j is an index referring to fibres where $L > L_\varepsilon$, η_0 is the fibre orientation factor, $V_{f,i}$ is the fibre volume fraction when i holds, and $V_{f,j}$ is the fibre volume fraction when j holds [26].

$$\tau = \frac{E_f D_f}{2L_{\varepsilon_c}} \tag{9.19}$$

$$\sigma_f = E_f \varepsilon_f \left(1 - \left(\frac{E_f \varepsilon_f D_f}{4L\tau}\right)\right) \tag{9.20}$$

$$\sigma_c = \eta_0 \left(\Sigma_i \left[\frac{\tau L_i V_{f,i}}{D_f} \right] + \Sigma_j \left[E_f \varepsilon_c V_{f,j} \left(1 - \frac{E_f \varepsilon_c D_f}{4L_j \tau} \right) \right] E_m \varepsilon_c (1 - V_f) \right) \tag{9.21}$$

9.4 Tensile failure at the macrostructural level

Different microstructural failures can be observed when composites break. Certain laminating arrangements favour some failure mechanisms over others, and in many instances, failure at the macrostructural level can be easily characterised by the abundance of a certain type of microstructural failure. Here, we will cover two typical modes of failure in tensile loaded continuous fibre-reinforced plastics: (1) cross-fracture and (2) splaying. Each of these failure modes is dependent on strength of the fibre–matrix interface. Figure 9.17 shows examples of each using carbon fibre/epoxy coupons tested in tension. Splaying is shown in figure 9.17(a), which occurs when the interface is weak. This can be a consequence of a semi-cured matrix, poor fibre dispersion, or fibre sizing-matrix incompatibility. Local brittle failure is shown in figure 9.17(b), which is a typical failure mode in composites with excellent fibre–matrix bond strength.

Environmental factors such as moisture ingress and age/heat-related hardening can also affect the strength of the fibre–matrix interface. Moisture diffusion into both thermoset and thermoplastic polymers is commonly modelled as Fickian. In composites, non-Fickian models are often deemed more appropriate since composites have inherently more complex microstructures [28]. In composite materials using Fick's second law of diffusion, moisture uptake, D, in immersed composites can be represented by equation (9.22), while the theoretical change in composite mass, M, is represented by equation (9.23). In these equations, M_1 and M_2 are the

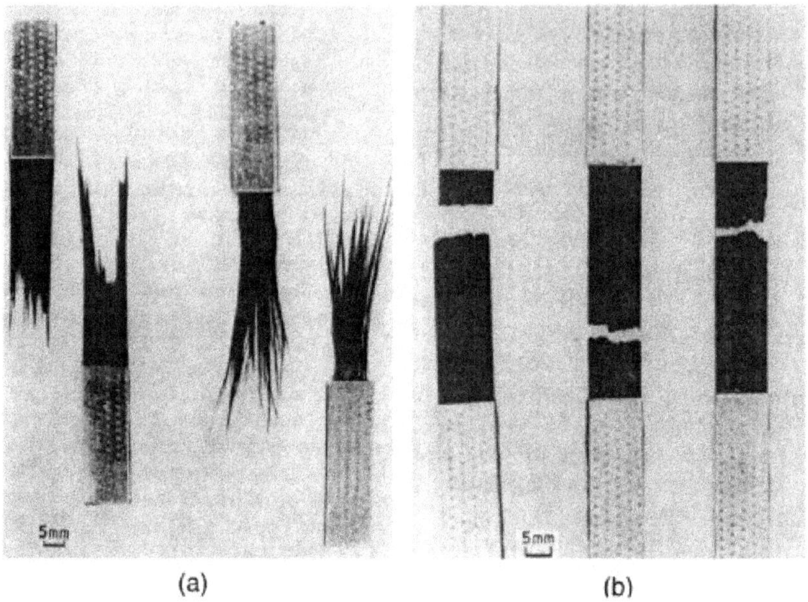

Figure 9.17. Fracture profiles for carbon fibre/epoxy composites in tension: (a) dispersed failure, or splaying due to a weak matrix and interface in shear, and (b) local brittle failure due to a strong matrix and interface. Reprinted from [27]. Copyright (2000), with permission from Elsevier.

moisture contents of the composite at times t_1 and t_2, respectively; sample thickness is represented by h; L is the sample length; w is the sample width; t is time; and M_∞ is the maximum change in total mass of the material. While this model is accurate to the point of M_∞, beyond this point, the model predicts a plateau, which is true in certain types of composite such as vinyl ester matrix composites [29] but is inaccurate in the case of polyester matrix composites where the mass change declines as a result of physical degradation within the composite, alongside potentially irreversible chemical degradation [30].

$$D = \pi\left(\frac{h}{4M_\infty}\right)^2\left(\frac{M_2 - M_1}{\sqrt{t_2} - \sqrt{t_1}}\right)^2\left(1 + \frac{h}{L} + \frac{h}{w}\right)^{-2} \tag{9.22}$$

$$M = \left[1 - \frac{8}{\pi^2}\exp\left(-\pi^2\frac{Dt}{h^2}\right)\right]M_\infty \tag{9.23}$$

Four main parameters affect the rate of water diffusion in fibre-polymer matrix composite. These are:

1. The rate of water diffusion through the bulk of the matrix material
2. The rate of water diffusion through the polymer interphase region
3. The interface between fibre and matrix
4. The fibre and its properties

In addition to these, factors such as fibre sizing, surface roughness, and defects will further create variations in the rate and style of water ingress. Nevertheless, in all cases the two main mechanisms of water ingress are diffusion and capillarity. Mechanisms of water diffusion will fundamentally be governed by molecular structure and their arrangements. Capillarity is generally confined to voids, defects, and at fibre–matrix interfaces where the water can wick.

As mentioned previously, heat will also impact the properties and behaviour of composite interfaces. For example, the interfacial responses of epoxy resins are linked to their glass transition temperature [31] as stress transfer between fibre and matrix is dissociated at temperatures both above and below the matrix Tg. A majority of studies on this topic note that over the matrix Tg, the fibre–matrix interface strength is hidden, as matrix flows in its rubbery state, which dominates composite behaviour. Above the Tg, interfacial debonding is often noted to be a primary failure mode [32–34], whereas below the Tg, there is a higher probability of interply failure [31, 32, 35]. A decrease in fibre–matrix interfacial strength is also enabled as a function of temperature-dependent variations in matrix fracture strains [32]. Furthermore, heating a composite excites molecular mobility at fibre–matrix interfaces, which destabilises secondary interactions at the interfaces and weakens the composite. Differential thermally induced expansions between fibre and matrix add interfacial stresses, consequently lowering the load-bearing capacity of a composite at failure.

Both water and heat conditioning are used conjointly to hygrothermally age composites as this approach maximises the rate at which composite interfaces (and components) degrade. Hygrothermal ageing in essence involves the water conditioning of composite material under elevated temperatures. These temperatures typically do not exceed the Tg of the material as doing so may activate oxidation reactions, altering the chemistry and properties of the materials, making it harder to conduct appropriate comparative studies. The combined effects of heat and water ingress through hygrothermal conditioning can be clearly seen in figures 9.18 and 9.19 where greater levels of splaying occur at higher hygrothermal ageing temperatures in Bouligand structured and discontinuous Bouligand structured carbon fibre-reinforced composites, respectively.

9.5 Compressive modes of failure in composites

Composites are non-homogenous materials, and as such, a single compression test will typically exhibit a range of different failure modes at the microstructural level. Certain failure modes may predominate and the more common modes of failure are covered in detail by Fleck [38], who considered the predominating failure modes as:

- Elastic microbuckling
- Plastic microbuckling
- Fibre crushing
- Splitting
- Buckle delamination
- Shear band formation

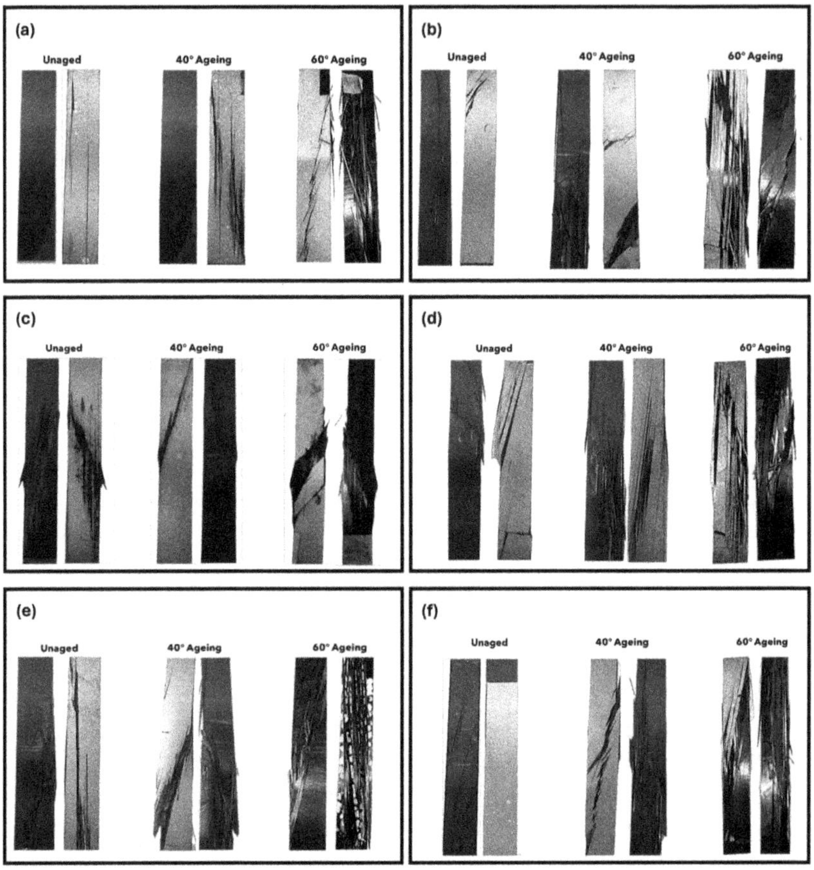

Figure 9.18. Fracture profiles for Bouligand-structured carbon fiber/epoxy composites in tension under unaged and 40 °C and 60 °C hygrothermally aged conditions: (a) 5° pitch angle, (b) 10° pitch angle, (c) 15° pitch angle, (d) 20° pitch angle, (e) 25° pitch angle, and (e) 30° pitch angle. Reproduced from [36]. CC BY 4.0.

Figures 9.20(a)–(f) provides schematics for each of the different microstructural failure modes in compression.

Elastic microbuckling is due to shear buckling instabilities. When fibres encounter shear buckling the matrix surrounding the reinforcing fibres will fail through pure shearing. The classical work of Rosen [39] assumed that fibres within the composite start as perfectly aligned and elastically deflect sinusoidally. Compressive strength of the composite, σ_c, is calculated according to equation (9.24), where E is the axial compressive modulus of the composite, G is the in-plane shear modulus, d is the fibre diameter, and γ is the sinusoidally shaped buckling wavelength. Here, G is the matrix component's elastic resistance to shear, while $\frac{\pi^2}{3}(\frac{d}{\gamma})^2E$ is the fibre component's elastic resistance to buckling.

$$\sigma_c = G + \frac{\pi^2}{3}\left(\frac{d}{\gamma}\right)^2 E \tag{9.24}$$

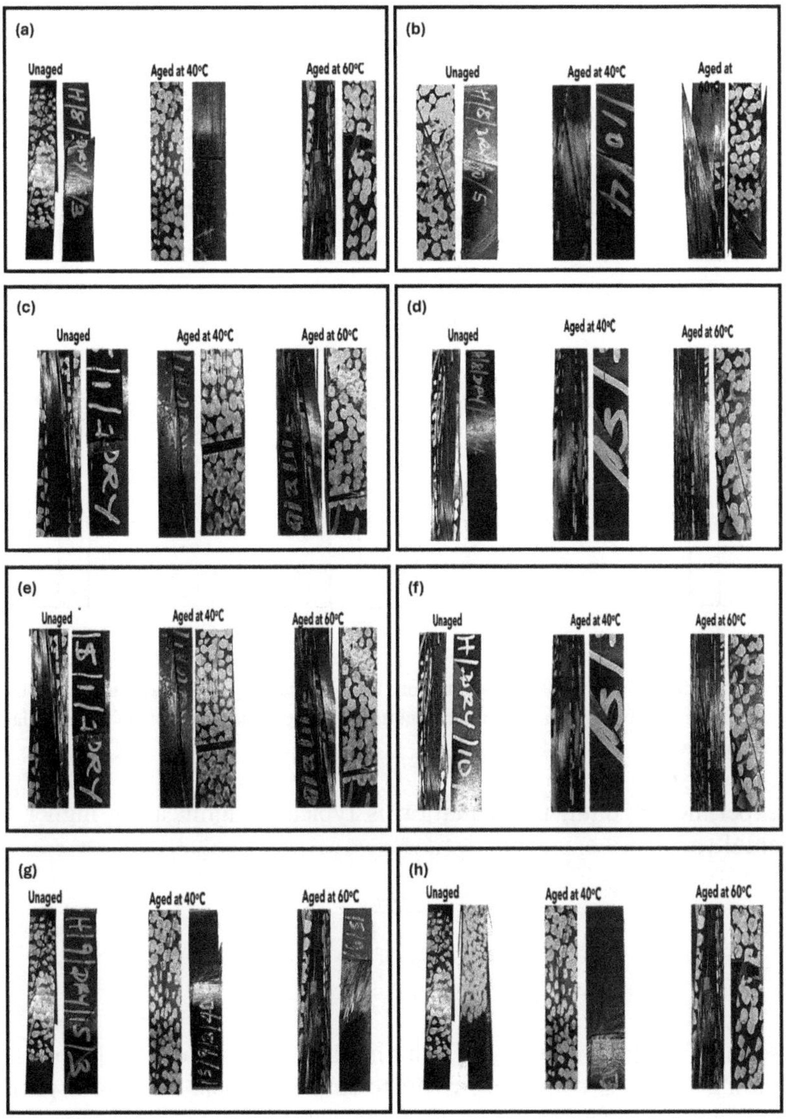

Figure 9.19. Fracture profiles for discontinuous Bouligand-structured carbon fiber/epoxy composites in tension under unaged and 40 °C and 60 °C hygrothermally aged conditions: (a) 5:90°, pitch angle, (b) 10:90° pitch angle, (c) 15:90° pitch angle, (d) 25:90° pitch angle, (e) 5:120° pitch angle (e) 10:120° pitch angle, (f) 15:120° pitch angle, and (g) 25:120° pitch angle. Reproduced from [37]. CC BY 4.0.

When the fibres in a composite are stiff, the Rosen equation can also be expressed according to equation (9.25), where G_m is the shear modulus of the matrix and V_f is the volume fraction of the fibre component in the composite. For cases where the fibre has an infinite shear modulus, this equation is expressed as equation (9.26), where $G_{13,c}$ is the shear modulus of the composite over the $_{13}$ plane. When assessed

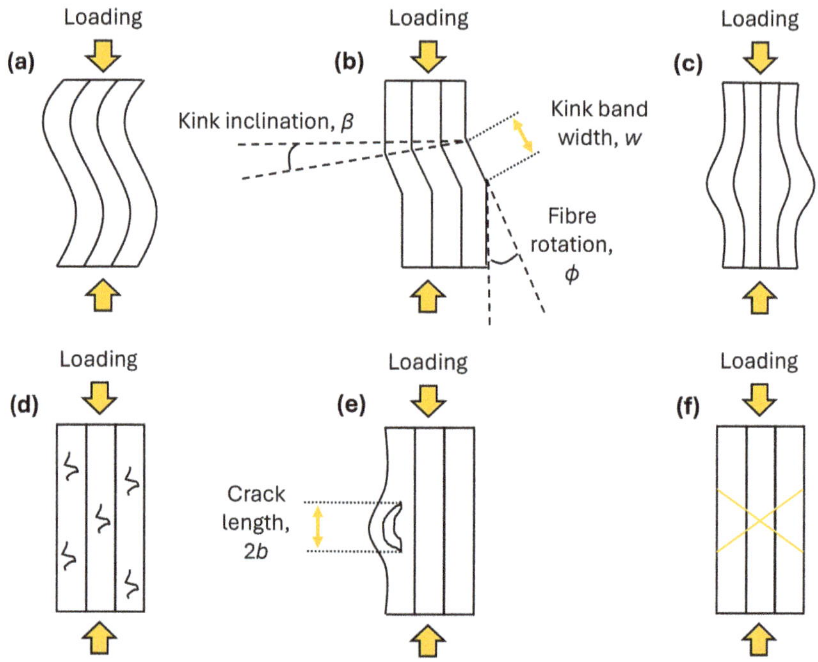

Figure 9.20. Competing modes of failure in continuous fibre-reinforced plastics: (a) elastic microbuckling, (b) plastic microbuckling, (c) fibre crushing, (d) matrix splitting, (e) buckle delamination of a surface layer, and (f) shear band formation.

against experimental data, this equation is typically significantly higher than for experimental values.

$$\sigma_c = \frac{G_m}{1 - V_f} \tag{9.25}$$

$$\sigma_c = G_{13,c} \tag{9.26}$$

In response to that $\sigma_c \neq G_{13,c}$, Tomblin and co-workers [17] discuss elastic microbuckling in terms of its sensitivity to imperfections, and more particularly, fibre misalignment. They postulate that the detection of imperfection sensitivity requires the incorporation of an elastic non-linear shear response into the original Rosen [16] formula. Chung and Weitsman [18] proposed alternative models based on geometrical imperfections such as fibre waviness and further took into account the randomness of fibre to fibre spacing, which they saw as an enabler for transitions between elastic microbuckling and plastic microbuckling, or microkinking.

Plastic microbuckling (microkinking) similarly to elastic microbuckling occurs due to shear instabilities. However, plastic microbuckling occurs at sufficiently high strains such that the matrix component of the composite deforms plastically. Polymer matrix composites that plastically microbuckle at failure typically have a compressive strength, $\sigma_c \approx \frac{3}{5}\sigma_t$, where σ_t is the tensile strength of the composite. The shear yield strength of the composite, τ_y, and the starting fibre misalignment angle, ϕ_0, are primary contributors to σ_c, equation (9.27) [40]. As the composite is deformed, the fibre angle ϕ can vary further and σ_c is then updated according to equation (9.28). In an elastic-perfectly plastic composite, equation (9.28) can be modified to include the yield strain, $\gamma_y = \frac{\tau_y}{G}$, as shown in equation (9.29) [41].

$$\sigma_c = \frac{\tau_y}{\phi_0} \tag{9.27}$$

$$\sigma_c = \frac{\tau_y}{\phi_0 + \phi} \tag{9.28}$$

$$\sigma_c = \frac{\tau_y}{\gamma_y + \phi_0} = \frac{G}{1 + \dfrac{\phi_0}{\gamma_y}} \tag{9.29}$$

A micromechanics model for kink band angle (for both long and short fibre waves), β, suggested by Budiansky [41] and further confirmed by Chaudhuri [45], is described by equation (9.30), where $\sigma_{1,c}$ is the axial composite compressive strength and $E_{3,c}$ is the composite compressive modulus in the 3 axis. Experimental evidence currently suggests that kink band angles are typically observed between 20° and 40° relative to the fibre axis [42–44]. However, as highlighted by Niu and Talreja [46], reliable theory to predict kink band angle has been sparse in the literature. When the elastic strain of the matrix is neglected, and the fibre is transversely inextensible, in an initial kink state ($\phi = 0$), the average kink angle is addressed according to equation (9.31), where H_m is the transverse to fibre axis height of the matrix component; H_f is the transverse to fibre axis height of the fibre component and ω is the slip line angle, which from Tresca is $\sin(2\omega) = \frac{2\tau_m}{\sigma_{y,m}}$; and τ_m is the matrix shear strength, while $\sigma_{y,m}$ is the matrix yield strength. According to the Tresca criterion, if $\omega = 45°$, then $\sigma_{y,m}$ reaches a minimum. Niu and Talreja nevertheless [46] hypothesise that equation (9.31) does not involve any post-kinking process as the assumption for β is initial kinking. As such, they suggest that for perfectly straight fibres embedded into a plastic matrix, $\omega = 45°$, and the kink band can be calculated in accordance with equation (9.32), where V_m is the matrix fraction.

$$\sigma_{1,c} = G_{13,c} - E_{3,c}\tan^2\beta \tag{9.30}$$

$$\tan\beta = \frac{2H_m}{H_f + 2H_m}\tan\omega = (1 - V_f)\tan\omega \tag{9.31}$$

$$\beta = \arctan(1 - V_f) = \arctan V_m \tag{9.32}$$

Fibre crushing is a consequence of shear instabilities within reinforcing fibres, such that buckling failure occurs within the fibres and leads to the fibres crushing within the composite under compression. As described by Fleck [38], it would seem that the fibre crushing strength, $\sigma_{f,c}$, in brittle or near-brittle composites, can be adequently predicted by equation (9.33), where V_f is the fibre volume fraction, V_m is the matrix volume fraction, E_f is the compressive modulus of the fibre, E_m is the compressive modulus of the matrix matter, and $\varepsilon_{f,c}$ is the strain at which the fibres fail in compression.

$$\sigma_{f,c} = (V_f E_f + V_m E_m)\varepsilon_{f,c} \tag{9.33}$$

Composite splitting is a characteristic failure mode of composites where both reinforcement and matrix exhibit brittle or near-brittle load-deformation behaviour. It is a dominant failure mode when matrix stiffness exceeds that of the reinforcement, which is more commonly the case in ceramic matrix composites. Mode I cracks develop in the fibre axis along voids, discontinuities, foreign inclusions, and other flaws. These act as local stress concentrators from which microcracks grow under progressively increased loading. At a critical load, the microcracks interact, resulting in failure along macroscopic shear bands. The compression strength, σ_c, of composites that fail predominantly in this manner can be calculated according to equation (9.34) where C is a coefficient that is expressed as a function of porosity p, a is crack length, and K_{IC} is the mode I critical fracture toughness of the matrix.

$$\sigma_c = \frac{C(p)K_{\mathrm{IC}}}{\sqrt{\pi a}} \tag{9.34}$$

Buckle delamination is an interfacial delamination that occurs due to local buckling of a surface layer from a subsurface debond. This type of delamination can be correlated to the coupling of low matrix toughness and the existence of large subsurface flaws. At its maximum, the compression strength, σ_c, can be represented by equation (9.35), where E_p is the plane strain value of the compression modulus, σ_0 indicates the onset of buckling where $\sigma_0 = \frac{\pi^2}{12}E_p(\frac{h}{b})^2$, b is the half crack length lying parallel to the free surface, h is the depth at which this crack exists below the surface, and G_c is the interfacial toughness.

$$\sigma_c = \sigma_0\left[1 + \frac{1}{2}\left(\frac{E_p G_c}{\sigma_0^2 h}\right)\right] \tag{9.35}$$

The last failure mechanism described by Fleck [38] is shear band formation, which is essentially the yielding and fracture of matrix material at a 45° orientation to the loading axis. This type of failure is often noted when composites are of low reinforcing volume fraction. When the reinforcing volume fraction decreases, the compressive strength $\sigma_c \rightarrow a\tau_m$, where τ_m is the shear strength of the matrix and a is a value that typically lies between 0.15 and 0.35. The Tresca criterion lies close to the centre of this range as it approximates the shear strength, τ, of isotropic materials as $\tau = \frac{1}{2}\sigma_c$.

The above are essentially competing failure modes and plastic microbuckling dominates over elastic microbuckling when equation (9.36) is satisfied, fibre crushing dominates plastic micrbuckling when equation (9.37) is satisfied, and splitting dominates when equation (9.38) is satisfied.

$$\frac{\phi_0}{\gamma_y} > 0 \tag{9.36}$$

$$\frac{\varepsilon_{f,c}E}{G} < \frac{1}{1+\left(\dfrac{\phi_0}{\gamma_y}\right)} \tag{9.37}$$

$$\frac{CK_{\mathrm{IC}}}{G\sqrt{\pi a}} < \frac{1}{1+\left(\dfrac{\phi_0}{\gamma_y}\right)} \tag{9.38}$$

9.6 Interlaminar shear failure in composites

Interlaminar shearing is a common failure mode in laminated composites. Table 9.1 provides details on common standards that have been used to determine the shear and interlaminar shear properties of fibre-reinforced composites. Standard methods for short beam testing, double beam shear testing, v-notched beam testing, v-notched rail testing, notched tensile testing, notched compression testing, and sandwich core shear testing are provided in this table. Schematic examples of different methods are shown in figures 9.21(a)–(d), where (a) shows a short beam shear test in 3-point bending, e.g. ASTM D2344, ISO 14130, BS EN 2377, BS EN 2563 [47–50]; (b) shows a v-notched shear test, e.g. ASTM D5379 [52]; (c) shows a double notched shear test; e.g. ASTM D3846 (note: withdrawn in 2024), BS EN 13121-3 [54, 55]; and (d) shows a sandwich core shear test, e.g. ASTM C393/C393M-20 [57] (figure 9.22).

Short beam bending is perhaps the most commonly used method for measuring interlaminar shear strength (ILSS). The ILSS is calculated using equation (9.39),

Table 9.1. Standard test methods for the measurement of shear and interlaminar shear in composites.

Standard	Type	Source
ASTM D2344, ISO 14130, BS EN 2377, BS EN 2563	Short beam test	[47–50]
ISO 19927	Double beam shear test	[51]
ASTM D5379	V-notched beam test	[52]
ASTM D7078	V-notched rail test	[53]
ASTM D3846 (note: withdrawn in 2024), BS EN 13121-3	Notched tensile test	[54, 55]
BS 6464	Notched compression test	[56]
ASTM C393/C393M-20	Core shear properties of sandwich constructions	[57]

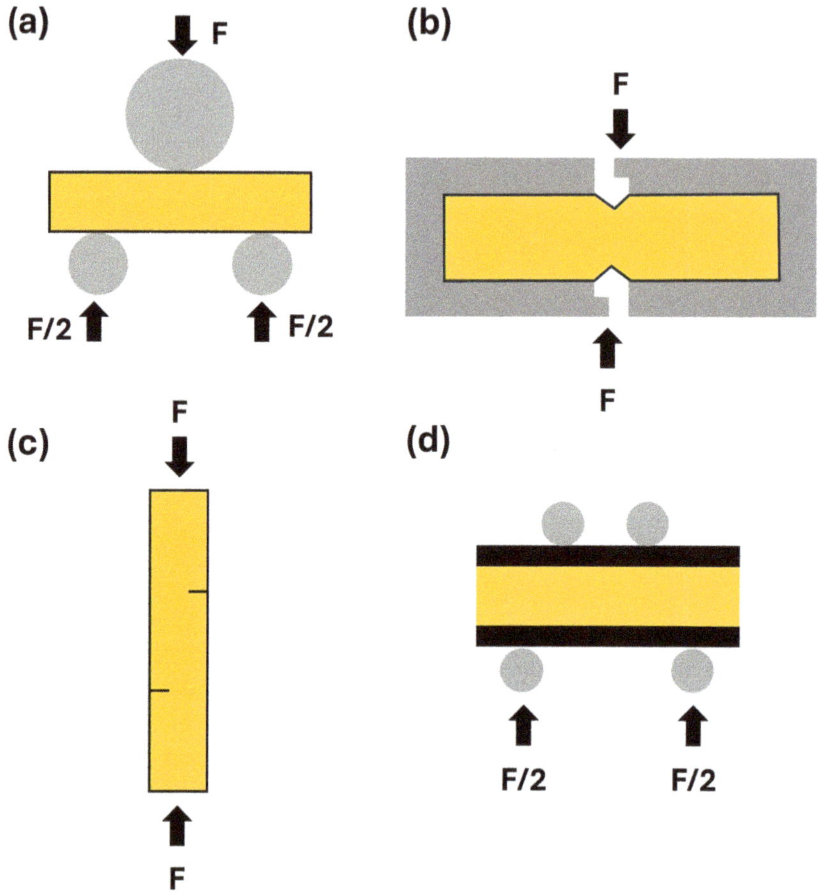

Figure 9.21. Schematic examples of methods for interlaminar shear testing (a) short beam shear test in 3-point bending, e.g. ASTM D2344, ISO 14130, BS EN 2377, BS EN 2563; (b) v-notched shear test, e.g. ASTM D5379; (c) double notched shear, e.g. ASTM D3846 (note: withdrawn in 2024), BS EN 13121-3; and (d) sandwich core shear test, e.g. ASTM C393/C393M-20.

where P is the imposed load and b and h are the width and thickness of the coupon, respectively.

$$\text{ILSS} = \frac{3}{4}\frac{P}{bh} \tag{9.39}$$

As described by Park and Seo [58], the ILSS can be calculated according to equation (9.40), where $G_{13,c}$ is the principal shear modulus of the composite, $_c$, with subscripts $_1$ and $_3$, referring to the fibre and transverse directions, respectively; Δu is the difference between the longitudinal displacement of the centres of two adjacent fibres in the through-thickness direction of the composite; s is the distance between the two adjacent fibres; and r is the average radius of a fibre.

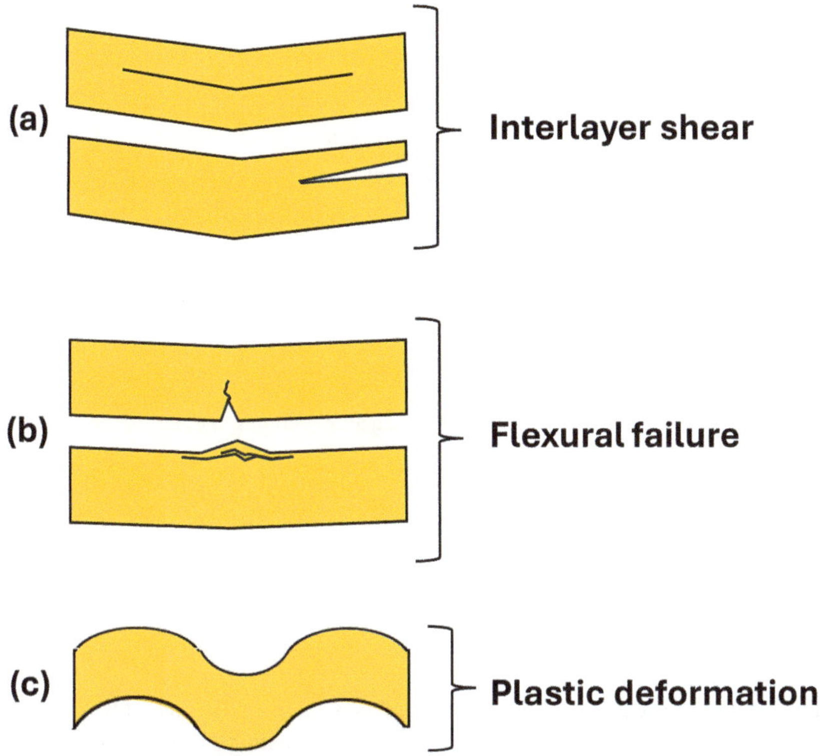

Figure 9.22. Examples of common failure modes observed in short beam tests (e.g. ASTM D2344, ISO 14130, BS EN 2377, BS EN 2563) specifically (a) interlayer shear failure, (b) flexural failure, and (c) plastic deformation of the short beam under rollers.

$$\text{ILSS} = G_{13,c}\frac{\Delta u}{s + 2r} \tag{9.40}$$

The maximum shear stress in the matrix, $\tau_{\max,m}$, is calculated according to equation (9.41), where G_m is the shear modulus of the matrix material.

$$\tau_{\max,m} = G_m\frac{\Delta u}{s} \tag{9.41}$$

The maximum shear stress from an ILSS short beam test, $\tau_{\max,\text{ILSS}}$, can be calculated by equation (9.42).

$$\tau_{\max,\text{ILSS}} = \text{ILSS} \cdot \left(\frac{G_m}{G_{13,c}} \times \frac{s + 2r}{s}\right) \tag{9.42}$$

If $\tau_{\max,m}$ is greater than the interfacial shear stress, $\tau_{\max,\text{ILSS}}$ gives an interface shear strength, whereas, if $\tau_{\max,m}$ is lower than the interfacial shear stress, $\tau_{\max,\text{ILSS}}$ is that of the matrix shear strength. This is an important distinction and should be taken into account when conducting ILSS short beam tests.

References

[1] ASTM D3822/D3822M-14 2020 *Standard Test Method for Tensile Properties of Single Textile Fibers* (West Conshohocken, PA: ASTM International)

[2] ISO 11 566:1996 *Carbon Fibre: Determination of the Tensile Properties of Single-Filament Specimens* (Geneva: International Organization for Standardization)

[3] ASTM C1557-20 2020 *Standard Test Method for Tensile Strength and Young's Modulus of Fibers* (West Conshohocken, PA: ASTM International)

[4] ASTM D2343-17 2023 *Standard Test Method for Tensile Properties of Glass Fiber Strands, Yarns, and Rovings Used in Reinforced Plastics* (West Conshohocken, PA: ASTM International)

[5] Hearle J W S 2002 Forms of fibre fracture *Fibre Fracture* ed M Elices and J Llorca (Oxford: Elsevier Science)

[6] Buehler M J 2006 Nature designs tough collagen: explaining the nanostructure of collagen fibrils *Proc. Natl. Acad. Sci. USA* **103** 12285–90

[7] Hosseinnezhad R, Elumalai D and Vozniak I 2023 Approaches to control crazing deformation of PHA-based biopolymeric blends *Polymers* **15** 4234

[8] Hu J, Deng X, Zhang X, Wang W-X and Matsubara T 2021 Effect of off-axis ply on tensile properties of [0/θ]ns thin Ply laminates by experiments and numerical method *Polymers* **13** 1809

[9] Sun C T 2000 Strength analysis of unidirectional composites and laminates *Comprehensive Composite Materials* **vol 1** ed A Kelly and C Zweben (Amsterdam: Elsevier B.V) pp 641–66

[10] Sudhir A and Talreja R 2019 Simulation of manufacturing induced fiber clustering and matrix voids and their effect on transverse crack formation in unidirectional composites *Composites A* **127** 105620

[11] Thongchom C, Refahati N, Roodgar Saffari P, Roudgar Saffari P, Niyaraki M N, Sirimontree S and Keawsawasvong S 2022 An experimental study on the effect of nanomaterials and fibers on the mechanical properties of polymer composites *Buildings* **12** 7

[12] Beura S, Chakraverty A P, Thatoi D N, Mohanty U K and Mohapatra M 2021 Failure modes in GFRP composites assessed with the aid of SEM fractographs *Mater. Today: Proc.* **41** 172–9

[13] Girard H, Koohbor B, Doitrand A and Livingston R 2023 Experimental characterization of in-plane debonding at fiber-matrix interface using single glass macro fiber samples *Composites A* **171** 107573

[14] Wells J K and Beaumont P W R 1984 Debonding and pull-out processes in fibrous composites *Defence Technical Information Centre Report Number: PB85-163343* (Springfield, VA: National Technical Information Service)

[15] Hashin Z and Rosen B W 1964 The elastic moduli of fibre-reinforced materials *J. Appl. Mech.* **31** 223–32

[16] Rosen B W 1970 Thermomechanical properties of fibrous composites *Proc. R. Soc.* **319** 79–94

[17] Tomblin J S, Barbero E J and Godoy L A 1997 Imperfection sensitivity of fiber micro-buckling in elastic-nonlinear polymer-matrix composites *Int. J. Solids Struct.* **34** 1667–79

[18] Chung I and Weitsman Y 1994 A mechanics model for the compressive response of fiber reinforced composites *Int. J. Solids Struct.* **31** 2519–36

[19] Livingston R and Koohbor B 2022 Characterizing fiber-matrix debond and fiber interaction mechanisms by full-field measurements *Composites C* **7** 100229

[20] Khan T, Hafeez F and Umer R 2023 Repair of aerospace composite structures using liquid thermoplastic resin *Polymers* **15** 1377

[21] Shahroze R M, Ishak M R, Sapuan S M, Leman Z, Chandrasekar M and Asim M 2019 Effect of silica aerogel additive on mechanical properties of the sugar palm fiber-reinforced polyester composites *Int. J. Polym. Sci.* **2019** 3978047

[22] Kelly A and Tyson W R 1965 Tensile properties of fibre-reinforced metals: copper/tungsten and copper/molybdenum *J. Mech. Phys. Solids* **13** 329–50

[23] Drzal L T and Madhukar M 1993 Fibre-matrix adhesion and its relationship to composite mechanical properties *J. Mater. Sci.* **28** 569–610

[24] Bowyer W H and Bader M G 1972 On the re-inforcement of thermoplastics by imperfectly aligned discontinuous fibres *J. Mater. Sci.* **7** 1315–21

[25] Bader M G and Bowyer W H 1973 An improved method of production for high strength fibre-reinforced thermoplastics *Composites* **4** 150–6

[26] Aliotta L and Lazzeri A 2020 A proposal to modify the Kelly-Tyson equation to calculate the interfacial shear strength (IFSS) of composites with low aspect ratio fibers *Compos. Sci. Technol.* **186** 107920

[27] Phoenix S L and Beyerlein I J 2000 Statistical strength theory for fibrous composite materials *Reference Module in Materials Science and Materials Engineering, Comprehensive Composite Materials* **vol 1** (Amsterdam: Elsevier) ch 1.19 pp 559–639

[28] Alam P, Robert C and Ó Brádaigh C M 2018 Tidal turbine blade composites—a review on the effects of hygrothermal aging on the properties of CFRP *Composites B* **149** 248–59

[29] Kootsookos A and Mouritz A P 2004 Seawater durability of glass- and carbon-polymer composites *Compos. Sci. Technol.* **64** 1503–11

[30] Liao K, Schultheisz C R, Huntson D L and Brinson L C 1998 Long term durability of fibre reinforced polymer matrix materials for infrastructure applications: a review *J. Adv. Mater.* **30** 3–40

[31] Kim Y J, Siriwardanage T, Hmidan A and Seao J 2014 Material characteristics and residual bond properties of organic and inorganic resins for CFRP composites in thermal exposure *Constr. Build. Mater.* **50** 631–41

[32] Peters P W M and Andersen S I 1988 The influence of matrix fracture strain and interface strength on cross ply cracking in CFRP in the temperature range of −100 °C to +100 °C *J. Compos. Mater.* **23** 944–60

[33] Miyano Y, Nakada M, Kudoh H and Muki R 1999 Prediction of tensile fatigue life under temperature environment for unidirectional CFRP *Adv. Compos. Mater.* **8** 235–46

[34] Miyano Y and Nakada M 2012 Formulation of time- and temperature- dependent strength of unidirectional carbon fiber reinforced plastics *J. Compos. Mater.* **47** 1897–906

[35] Cao S, Wu Z and Wang X 2009 Tensile properties of CFRP and hybrid FRP composites at elevated temperatures *J. Compos. Mater.* **43** 315–30

[36] Nwambu C, Robert C and Alam P 2022 The tensile behaviour of unaged and hygro-thermally aged asymmetric helicoidally stacked CFRP composites *J. Compos. Sci.* **6** 137

[37] Nwambu C, Robert C and Alam P 2022 Tensile behaviour of unaged and hygrothermally aged discontinuous Bouligand structured CFRP composites *Oxford Open Mater. Sci.* **3** itac016

[38] Fleck N A 1997 Compression failure of fiber composites *Adv. Appl. Mech.* **33** 43–117

[39] Rosen B W 1965 Fiber composite materials *Mechanics of Composite Strengthening* (Materials Park, OH: American Society of Metals) ch 3 pp 37–75

[40] Argon A S 1972 Fracture of composites *Treatise on Material Science and Technology* **vol 1** ed H Herman (New York: Academic) pp 79–114
[41] Budiansky B 1983 Micromechanics *Comput. Struct.* **16** 3–12
[42] Skovsgaard S P H and Jensen H M 2019 A general approach for the study of kink band broadening in fibre composites and layered materials *Eur. J. Mech. A* **74** 394–402
[43] Patel J, Ayyar A and Peralta P 2020 Kink band evolution in polymer matrix composites under bending: a digital image correlation study *J. Reinf. Plast. Compos.* **39** 852–66
[44] Budiansky B, Fleck N A and Amazigo J C 1998 On kink band propagation in fiber composites *J. Mech. Phys. Solids* **46** 1637–53
[45] Chadhuri R A 1991 Prediction of the compressive strength of thick-section advanced composite laminates *J. Compos. Mater.* **25** 1244–76
[46] Niu K and Talreja R 1998 Modeling of compressive failure in fiber reinforced composites *Int. J. Solids Struct.* **37** 2405–28
[47] ASTM D2344/D2344M-16 2016 *Standard Test Method for Short-Beam Strength of Polymer Matrix Composite Materials and Their Laminates* (West Conshohocken, PA: ASTM International)
[48] ISO 14 130:1997 Fibre-reinforced plastic composites—Determination of apparent interlaminar shear strength by short-beam method, Technical Committee: ISO/TC 61/SC 13 Composites and reinforcement fibres
[49] BS EN 2377:1989 Specification for glass fibre reinforced plastics. Test method. Determination of apparent interlaminar shear strength, British Standards International
[50] BS EN 2563:1997 Carbon fibre reinforced plastics. Unidirectional laminates. Determination of the apparent interlaminar shear strength, British Standards International
[51] ISO 19 927:2018 Fibre-reinforced plastic composites—Determination of interlaminar strength and modulus by double beam shear test, Technical Committee: ISO/TC 61/SC 13 Composites and reinforcement fibres
[52] ASTM D5379/D5379M-19e1 2019 *Standard Test Method for Shear Properties of Composite Materials by the V-Notched Beam Method* (West Conshohocken, PA: ASTM International)
[53] ASTM D7078/D7078M-20e1 2020 *Standard Test Method for Shear Properties of Composite Materials by V-Notched Rail Shear Method* (West Conshohocken, PA: ASTM International)
[54] ASTM D3846-08 2015 *Standard Test Method for In-Plane Shear Strength of Reinforced Plastics* (West Conshohocken, PA: ASTM International)
[55] BS EN 13 121-1:2021 GRP tanks and vessels for use above ground—Raw materials. Specification conditions and acceptance criteria, British Standards International
[56] BS 6464:1984 Specification for reinforced plastics pipes, fittings and joints for process plants, British Standards International
[57] ASTM C393/C393M-20 2020 *Standard Test Method for Core Shear Properties of Sandwich Constructions by Beam Flexure* (West Conshohocken, PA: ASTM International)
[58] Park S J and Seo M K 2011 Modeling of fiber–matrix interface in composite materials *Interface Science and Technology* **vol 18** ed S J Park and M K Seo (Amsterdam: Elsevier) ch 9 pp 739–76

www.ingramcontent.com/pod-product-compliance
Ingram Content Group UK Ltd.
Pitfield, Milton Keynes, MK11 3LW, UK
UKHW051852100926
8556IPUK00005B/591